STELLAR SINGULARITY

Navigating the Spacefaring Economy

ELIZEBETH VARGHESE

Soul Excellence Publishing

Copyright © 2022 by Elizebeth Varghese

All rights reserved.

No part of this book may be reproduced in any form or by any electronic or mechanical means, including information storage and retrieval systems, without written permission from the author, except for the use of brief quotations in a book review.

Published by Soul Excellence Publishing

Paperback ISBN: 979-8-9856016-9-5

eBook ISBN: 979-8-9856016-8-8

First Edition

For Mathew, my best friend and collaborator;
And for Aanika and Lara, the meaning of life.

Praise for Elizebeth Varghese

"Elizebeth Varghese may not be the first space philosopher, but she certainly deserves the title. *Stellar Singularity* integrates the technological aspects of our expansion into space with imperatives for human behavior—particularly economic, diplomatic and ethical behavior. Knowledgeable people outside of the traditional space enterprise can and must contribute to this next phase of human development. Stellar Singularity is the work they should begin with to appreciate its challenges and opportunities."

—GORDON ROESLER, PH.D., PRESIDENT, ROBOTS IN SPACE LLC, AND FORMER DARPA SPACE ROBOTICS PROGRAM MANAGER

"I've been tracking New Space for a short while and it's good to read about the New World in space, notably the ethical and geopolitical considerations as we venture further together. The interstellar journey is not fully recognized yet, just as many people are unaware of the role space plays in their day-to-day lives.

This book is a must-read for those wanting to broaden their horizons to space and the stellar singularity, it provides a great summary of historical and technological developments through to shifts in democratization, policy and value chain realization. I'd recommend it to anyone with an enquiring mind."

—STEVE RAMAGE, CHIEF ENGAGEMENT OFFICER - GROUP ON EARTH OBSERVATIONS (GEO), GOVERNING BOARD -

Digital Earth Africa, Earth Observation Advisory Committee - UK Space Agency

"This timely, thoughtful book is the perfect guide and roadmap for business leaders who are trying to make sense of the new spacefaring age we have entered. In her spirited second book, futurist Elizebeth Vargheses brings clarity, concision, and a human-centric lens to this critical conversation."

—Scott Stalker, Command Senior Enlisted Leader at United States Space Command, Active Duty US Marine with over 30 years of experience, Senior Fellow Irregular Warfare Initiative

"The latest must-read book from futurist and humanist, Elizebeth Varghese! *Stellar Singularity* enriches our understanding of how technology and the meaning of life are converging in this new age."

—Dr. Robert Zubrin, President, Mars Society

"In this engaging, authoritative and highly readable book, Elizebeth Varghese takes us on a journey through the complexities and dynamics of the emerging 'spacefaring age,' seamlessly moving between topics as diverse as Artificial Intelligence, Moon Mining, and an 'interstellar Magna Carta,' to lay out what our space future may hold."

—Professor Andrew Futter, University of Leicester, UK

"Stellar Singularity hits the market at a pivotal point in the evolution of the global space economy, estimated to become a $ 1 - 3 trillion dollar economy by 2040. No longer are we talking about putting a man on the moon, but going back to the moon—to stay, to learn, and to live —and then we'll go on to Mars, and even beyond to other planets, moons and stars. No longer are we only looking out at space exploration, but we are looking back at how space can solve our greatest challenges here on Earth. Timely and inspirational, *Stellar Singularity* speaks to our mission at The Space Foundation of global collaboration, so together we can work toward the betterment of humanity. "

—SHELLI BRUNSWICK, FORMER CONGRESSIONAL LIAISON - U.S. AIR FORCE, COO OF THE SPACE FOUNDATION, UNITED NATIONS SPACE4WOMEN MENTOR

"This was a great read touching upon the concept of 'stellar singularity' in our presently fast-growing space economy, including a historical comparison and analysis of past singularities (i.e. tipping points) in human history. I also enjoyed Elizebeth's argumentation on the need for our spacefaring industry to think (very) differently about how to tackle the ground rules of space exploration with non-traditional leadership values, which she expands upon."

—ROBERT FEIERBACH, CEO 0-G LAUNCH, HYPERSONICS AND SATELLITE EXPERT

"This is an excellent, intelligent, and well-thought-out introduction to the issues revolving around the expansion of humans into outer space. Beginning with the history of exploration of the Western Hemisphere over 500 years ago, this book describes the possible similar trends for the future of space colonies.

This is a must-read for people interested in the future of space civilizations, and the issues and legal aspects that will follow. In addition, this is a current issue for moon settlements and ownership and resource "rights". Some have already talked about the potential renunciation of the Outer Space Treaty of 1967 by some countries who may be considering declaring total ownership on the moon, much like Spain claimed ownership of the entire Western Hemisphere after Christopher Columbus' explorations. These issues will certainly affect our children and us for a long time."

—Paul Szymanski, Space Strategies Center, International recognized speaker and author on the militarization of outer space

"As we witness the democratization of our access to space, we are at the cusp of a new paradigm of business, society, and global governance. Stellar Singularity makes a compelling argument that now is the time to think and prepare for a new age that holds ample promise for a bright future."

—Roland Deiser, Chairman, Center for the Future of Organization at the Drucker School of Management and author of Organizing for Business Ecosystem Leadership

"From quarks to atoms, tardigrades, shattered windows, the Magna Carta and octopi—this is a story of singularity where you cannot go back. Explosions of ideas and concepts that open your mind like the JWST has opened our view of the universe. Discussions of the New Space leader capability skill set of competence, integrity, clarity and empathy will bring to our changing world hope, common good and optimism for the future. Thought provoking and real. It's about the journey— not the destination."

—Jackie Carpenter, Managing Director at One Giant Leap Australia Foundation

"Fascinating! A revolutionary take on the future of space and technology, Elizebeth teaches companies and leaders how to keep pace with the innovation that will take us farther than we thought possible. Well researched and brilliantly explained, you won't want to put this book down!"

—Dr. Marshall Goldsmith, Thinkers50 #1 Executive Coach and New York Times bestselling author of *The Earned Life*, *Triggers*, and *What Got You Here Won't Get You There*

"No one doubts the incredible impact of technology on all aspects of our lives. Using space exploration as a case, Elizebeth highlights the risks and opportunities in singularity, when the future is out of control and unpredictable. Adapting lessons of stellar singularity, she cleverly highlights how to manage in a world of unprecedented uncertainty. This is not only a book about space singularity, but managing organizations in the new world of technological disruption."

—Dave Ulrich, Rensis Likert Professor, Ross School of Business, University of Michigan Partner, The RBL Group

"Navigating the economy of space requires an inspired imagination, yet must remain grounded by the gravity of ethics. I am excited to read how Elizebeth balances the philosophical quest for knowing the unknown beside the moral imperative of sharing Global Commons in space. *Stellar Singularity* aims to represent the full spectrum of humankind's needs in space, a most admirable pursuit."

—S. Sita Sonty, Global Aerospace Executive, former U.S. diplomat, SpaceX veteran

"A visionary and accurate assessment, Elizebeth Varghese's book *Stellar Singularity* paints a comprehensive picture of the emerging space industry, its past and present state, and enlightens us to a centergistic momentum that is bringing humanity closer to reaching the stars."

—Kevin DuPriest, CEO Planetoid Mines

"From Spanish-days' Columbus, old '60's space race, to current expansion in the global space exploration and participation era, the book is inspirational and thought-provoking. In the coming years, we will be progressing, working, and collaborating differently. Use of space towards a global commons will be happening, where keeping the health and sustainability of both the Earth and space will be collaborative efforts. In our personal struggles navigating the geopolitical issues in our own entrance to space exploration in the last few years, we found that whether our current differences might or might not exist or resolve in the upcoming space exploration era in due time, space technology will become more available, more issues will be encountered and resolved in global governance and frameworks. We have a right to live, survive, and thrive together. We will and should find a better way. Hence, *Stellar Singularity* discusses leadership role possibilities quite thoroughly.

This book makes participants like us think through the developmental history of the space economies, and so we can find our positions, roles, and possible contributions into the future."

—Kuang-Han Ke, Chairman Gran Systems

"Elizebeth succinctly summarizes the current situation in the space domain and outlines opportunities for continued dialogue between the civil, commercial, and military sectors as well as the international community. Space professionals and enthusiasts alike will gain a better understanding of the global space issues after this important read."

—Bill Woolf, Col (ret.), USAF, President and Founder, Space Force Association

"A thoughtful, personal, well-researched, and timely piece of our human space exploration efforts, as we have reached an inflection point with this journey, evolving to a more diverse, humanistic, inter-disciplinary, and sustainable approach towards human habitation of space. Read more about the space industry's history, cultural connections, influential socioeconomic factors, key technological drivers, and how space exploration benefits all, directly and indirectly, from Elizebeth's must-read latest work".

—Dr. S. Anand Narayanan, Research Professor, Florida State University, Expert in Space and Medical Physiology, Global Health, Interdisciplinary STEAM Research and Education, NASA Fellow

"This interesting and insightful book takes you from a place of thinking about spaceflight as something that only rocket scientists engage in to a place of thinking about space holistically as a part of the new economy for the next century. As a business leader and a rocket scientist, I found the concepts particularly thought provoking as they pertain to the type of leaders we will need for all business in the coming decades."

—Mukwatsibwoha Alibaruho, Senior Vice President, Mission Launch and Recovery Operations, Sierra Space

"A Grand Tour of the issues, options, promises, and challenges! Elizebeth touches every side of the multifaceted, and daunting, notion of a spacefaring economy. Technological advances and advantages that promise to cure the many ills here on Earth along with the inevitable and unfortunate export of our geopolitical differences off-planet. Elizebeth offers many historical tools and policies that could offer hope in space while noting that current behavior and trends dampens that hope. Underlying all of it is the realization that, while everything we have learned over millennia about international cooperation on Earth offers many lessons, we are attempting to upload it all in a single push to interplanetary space. A fascinating read!"

—Dan Geraci, President, Voyager Space Services, Chairman, The Planetary Society

"*Stellar Singularity* goes beyond the technology to remind us that space is a social science and it is how humans interact with and in space that will determine our collective destiny. Elizebeth illustrates how the new space economy breaks down the old barriers to access and democratizes the use of space in a changing world. While Varghese allows us to enjoy the wonder of exploration, she reminds us that history provides a cautionary tale as we expand into new worlds. Her work brings to the forefront important ideas about the interaction of social disciplines in the global space society in an approachable and provocative way."

—Dr. Ruth Stilwell FRAeS, Executive Director Aerospace Policy Solutions, adjunct faculty

"*Stellar Singularity* synthesizes a rather large swath of thinking about space from a variety of fields, and asks some big questions and poses some grand ideas. I'm happy to see this type of broad discussion of space activities, especially the placing of them in a larger historical context."

—Christopher Johnson, Space Law Advisor, Secure World Foundation

"The book *Stellar Singularity: Navigating the Spacefaring Economy* by Elizebeth Varghese is a must-read for all interested in understanding the relevant challenges and opportunities of the new space age. The book's chapters masterfully present the most pertinent topics, such as technology, public-private partnerships, the meaning of what is to be human, new rules of engagement, leadership, and other significant considerations. From the beginning, Elizebeth Varghese reminds us of the human imperative to learn and explore. These thoughts evoke the last two Apollo missions. Their missions' logos suggest the first age of exploration and humanity's entrance into the solar system. But suppose humanity must enter a new age of discovery. In that case, it must learn to incorporate new rules of engagement with a greater understanding of the difficulties associated with the new cosmic terrain.

Elizebeth Varghese takes a unique and pragmatic approach, sharing memorable stories of humanity while searching for answers. She skillfully, and with tones of scholarly poetry, lets us appreciate the human condition. The author discusses the various subjects with a practical approach that considers the evolving nature of new technologies.

The unprecedented nature of the new space age means that new rules are necessary to provide a comprehensive approach supported by recognized responsibilities. Educators, entrepreneurs, government officials, students, parents, and the general public will find this book engaging and thought-provoking. The content of this book is not merely about space exploration; it is, in essence, a call to awareness and further action at the intersection of the use and exploration of outer space, cooperation, governance, and leadership in search of the 'Stellar Singularity.' Elizebeth Varghese has become one of the leading authorities on the subject of 'new space,' and her book will undoubtedly offer a valuable vehicle for its understanding."

—ROY BALLESTE, DIRECTOR AND PROFESSOR OF LAW, STETSON UNIVERSITY COLLEGE OF LAW

"When we learn something new, our perception of the world changes forever. The journey you take through the knowledge and discussions presented in *Stellar Singularity* will have this power of altering the reality around you. This book from futurist and humanist Elizebeth Varghese will open your mind to a new space era full of challenging situations and different perspectives. You will never be the same again—this is guaranteed."

—Professor Thais Russomano, MD, PhD – CEO InnovaSpace

"In a world increasingly dominated by nationalistic politics, it is a real treat to read a book that explores the nascent but growing global cooperation in areas like space technology and global governance and agreements regarding the use of outer space. S*tellar Singularity: Navigating the Spacefaring Economy* is a terrific lens into the acceleration of space technology, the increased participation and activity in outer space exploration, and the ongoing efforts and collaboration in the 'new' economy above the earth's orbit. The implications for global leadership and enhanced global cooperation is extraordinary. This is really a first of a kind book and a good read at that!"

—Mark Arian, CEO Korn Ferry Consulting

"*Stellar Singularity* provides a compelling and provocative reason why we should all care about space exploration and what it means for all of us on this planet. Beginning with Columbus and moving into the world of the ethical choices technology is requiring us to make, Varghese will make you a convert into understanding why if we care about the future of our planet, we must begin today to establish the ethical norms and practices for navigating and occupying space."

—Audrey Zibelman, VP at X, The Moonshot Factory

"Elizebeth Varghese has provided a very broad observation of the transition from past/current Space endeavours to what the future holds. *Stellar Singularity* is a good treatise on observations from the humanities into what the amazing new technology that is rising daily to transform the current Space economy into something completely different. From the ideas of 'democratizing space' to the new geopolitical boundaries and alliances that will inevitably be formed in our movement off Earth, her book helps to tie it all together into a single thought provoking discussion for anyone to get up to speed on New Space!"

—David Barnhart, Research Professor and Director Space Engineering Research Center, University of Southern California, and founder/CEO of Arkisys Inc.

"Varghese's contribution to the education of the future Space Singularity will help humanity's understanding of the hope and realities of what the $4T dollar Space Economy will provide."

—George S Pullen, Chief Space Economist, MilkyWayEconomy

"The beauty of *Stellar Singularity* is that it connects all Humanity to our original roots. Not only are we stardust, Varghese outlines how we're finally headed back home."

—Samson Williams, Adjunct Professor, University of New Hampshire School of Law

"Varghese focuses our attention like a laser on what might be humanity's last opportunity to 'get it right' for the outer space environment. Only an author with Varghese's breadth of vision could write with such urgency about the final frontier, space. Varghese's winsome prose makes the arguments clear for us all to use this final frontier with care. We are all space-faring, and we all have the opportunity— and the obligation—to learn from past mistakes and explore this final frontier with care and humanity."

—RUTH PRITCHARD-KELLY, SENIOR ADVISOR, SATELLITE REGULATORY AFFAIRS & SUSTAINABLE SPACE POLICY" ONE WEB

"*Stellar Singularity* synthesizes a rather large swath of thinking about space from a variety of fields, and asks some big questions and poses some grand ideas. I'm happy to see this type of broad discussion of space activities, especially the placing of them in a larger historical context."

—CHRISTOPHER JOHNSON, SPACE LAW ADVISOR, SECURE WORLD FOUNDATION

"Elizebeth presents an exciting narrative tying exploration and industry of the past (and present) to humanity's current and future expansion into space. The 1960s were the start of our Space Age. We are now on the cusp of another explosion of activity and innovation for the 'stellar singularity.' Her book provides convincing arguments for now being the time to set the ethical and legal frameworks for healthy commercial exploration and mining in space. Because, as she says in the book, soon, 'all companies will be sSpace companies.'"

— DAVID M VERMILLION, THE SPACE DATA GUY

asato mā sadgamaya
tamasomā jyotir gamaya
mrityormāamritam gamaya[1]

From ignorance, lead me to truth
From darkness, lead me to light
From death, lead me to immortality

Acknowledgments

Mumbai in 1978 was an uncomplicated and joyous place. We had one black-and-white television channel, so there was never any debate about what to watch. The highlight was "Chayageet" (Hindi movie songs) which played on Thursdays at 8.30 pm for half an hour. A Bollywood movie aired each Sunday evening, and the city shut down. I didn't care much for the big screen and napped through the mega-blockbuster film "Amar Akbar Anthony"...

Then one day, we took a family trip to the Nehru planetarium. I sensed magic in the dark, looking up into the night sky while the stars and planets appeared above.

I couldn't stop talking about what we had seen. After enduring planetarium reenactment, my parents suggested that my kindergarten class should hear about the trip.

I stood in front of the room, as sixty, five-year-olds stared at me, scrunched my favorite red, smocked dress with both hands, and stumbled into my opening.

I forgot everything else. After all, nothing was more heartening than what lay in the skies.

Thank you, Amma and Acha, for that first glimpse of a universe of knowledge. And, for the innumerable copies of National Geographic.

I'm ever grateful to my husband Mathew for the encouragement. Special thanks to Anu Isaac, who helped with the title and design - sisters can always be counted on, and will tell it like it is! Thanks also to Ritendra Roy for reading and editing the manuscript. I thank Charul Jain for facilitating a scholarly review of the content on Jainism.

My sincere thanks to Gordon Roesler, President of Robots in Space, formerly of DARPA, for his detailed review and advice. I am very grateful to Mathew (also with one "T") Kaplan, Host and Producer of Planetary Radio at The Planetary Society for his feedback and edits. I thank Commander Scott Stalker of the US Space Force for his support. Thank you to all my distinguished and expert reviewers for the kindness, encouragement, and thoughtful feedback. I'm so humbled by the interest and collaboration - and remain inspired by the global community that is building the New Space era.

I'm very thankful to my publisher, Kayleigh O'Keefe, and the editorial team at Soul Excellence Publishing for their support.

Elizebeth Varghese
September 15, 2022
New York, New York.

Introduction

OUR STELLAR SINGULARITY

In the summer of 2022, our thirteen-year-old waited in the backseat of a rented car where the blazing Madrid sun toasted the seats. Lara watched, amused, as Mathew and I struggled with the dashboard. It took us a while to interpret the display screen with prompts and buttons only in Spanish. But finally, we figured out the keywords and guidelines, started the navigation, turned on the Bluetooth, and hit the road. We turned the volume up on the podcast and slid into the fifteenth century.

As we drove across the Iberian peninsula, we got deeper into the territory of Isabella of Castille. We grew fascinated by the history and stories of "The Reconquista."[1] We drove past the rolling hills of La Mancha, studded with gnarled olive trees in groves, interspersed with fragrant orchards of bright, plump oranges. We sometimes saw the countryside through a haze as the summer heat steamed the tarmac. Cervantes' knight seemed to be among the windmills - his mythical and passionate quest for valor lingering on the slopes. We also found ourselves using more

Introduction

Spanish words. We lost our diffidence and bravely reached out, loving the warmth of the Castilians.

We learned how to navigate; we found the language, maps, the guardrails to travel these new highways. And we found that our experience grew more significant and memorable as we went deeper and further.

Everything had become more exciting and possible because we understood the journey.

And now, it was late in the day as we drove into Cordoba. Quiet, cobbled streets, flowers, and somnolent heat steeped in history. The setting sun lit the walls of the Alcazar. I could imagine we were back in 1486, when Isabella and her advisors worked late into the day, endlessly debating the wisdom of funding the Genoese explorer, Christopher Columbus.

Should the state support the exploration of the Indies? Could it? What would the returns be? How far would they go? Was it affordable? How should they use commercial funding? Who would grant the rights of ownership of the new territories? How would they cement the claim to the New World they were hoping to discover?

Isabella, Ferdinand, and their advisors vacillated. For years. Should they do what France, England, and Portugal had not chosen to do?

We left Cordoba and swung by Seville, then looped back to Grenada to visit the most magical fortresses - the palatine

Introduction

of The Alhambra, where Isabella had ridden up in triumph after the Moors had fallen.

Columbus, giving up, rode away and traveled dejectedly alongside the Jewish population expelled from Spain.[2] But, flush from the long-fought victory over the Moors, the Catholic Monarchs and their advisors in the splendid Nasrid palace changed their minds and recalled him.

Six years after they met Columbus, Isabella and Ferdinand decided to invest and supplement the private investors in Italy. In exchange, Columbus would give the monarchs silk and precious goods from "the Indies."[3]

Spain quickly brokered two papal bulls to claim the discovered lands under the church's power.[4] Pope Alexander VI wrote in his Inter Caetera to the effect that… "…by the authority of Almighty God…we give, grant, and assign to you and your heirs and successors…together with all their dominions, cities, camps, places, and villages, and all rights, jurisdictions, and appurtenances, all islands and mainlands found and to be found, discovered and to be discovered towards the west and south…".

The pope bestowed upon Spain everything from the Arctic to the Antarctic pole and everything to be found in the direction of India, one hundred leagues (about three hundred miles) towards the west and south from any of the islands of the Azores and Cape Verde.

The pope also forbade anyone (under the penalty of excommunication) "to go for…trade or any other reason.."

Public-private partnerships, trans-national discussions, treaties (like the Treaty of Tordesillas in 1506)[5] to carve up the new world, decisions by committees, leaders who backed a vision, explorers who broke laws, and the invoca-

Introduction

tion of non-existent authority to cement claims. As it was in 1492, it remains 500 years later.

We don't know what we'll find on our journey into the skies. What planets will we discover? What will we do when we find out about them? Who is calling the shots? Who will make the rules? We are balancing not only human awe and wonder but also human greed. There are positives and negatives. There are possibilities both promising and dangerous. Columbus ended up being arrested by his successor Francisco de Bobadilla for his brutalities as a colonial governor. His successor returned him to Spain in chains.

We must keep the principles of good governance and ethics at the forefront as we navigate ahead so our journeys remain on the right side of history and knowledge. And we need scientific humility and discussion as we expand our knowledge daily.

My family's Spanish sojourn reminded me we're a little further on, beyond where Columbus was in time. But just a little. The first space explorations took off and returned successfully. We're rapidly finding more planets, stars, and galaxies than we ever imagined. And we're at a point where settlers plan to make their way into the New World. The race is on. A lot more awaits us in the skies above.

We are at the cusp of a tremendous new singularity, driven by our journey among the stars.

Our Stellar Singularity awaits.

Introduction

People often talk about technological singularity and the fact that technology accelerates exponentially through a very dark lens. The mood is doom and gloom—almost dystopian, regarding the blurring lines between humans and machines.  This perspective is inherently short-sighted and pessimistic and speaks to a lack of ability to create the correct distinctions and parameters as we see things evolve. I'm here to illuminate the possibilities.

Similarly, there is much consternation about the acceleration of technology and how it may blind us to distinctions between right and wrong. This stems from a lack of a clear rubric of how we understand what life is (another topic we'll tackle in this book). But we do have clarity on what is "free will," which allows life of biological and organic origin to make ethical decisions and balance self-protection and the common good. The importance of collaboration toward the common good and governance that benefits all is key to our journey as we enter the era of a spacefaring economy. Let's open our guidebooks, learn the language and build our navigation plan.

This book explores ways to adapt and adjust thoughtfully to a new realm of possibilities. For now, know that the current singularity we are journeying through is a function of our current state and reality—no different than countless other shifts that have played out in history. The only difference, in this case, is that more people have a voice and a view of what is happening in real time.

Introduction

We've talked about the challenges in how we perceive change. We've overcome what we perceive as doomsday thinking before. In the 1700s, the English priest Thomas Robert Malthus penned an essay called the "Essay on Population."[6] He essentially said that, at a certain point of population, the death of humans was sure to follow.

His thinking was that if we tried to sustain an overwhelming number of people on the resources from the same amount of land, we could not produce sufficient food to support the population. He said there was a point at which things would not work anymore; our resources would be stripped dry.

But, what happened? How are we all still here? Well, there were new developments in science. We discovered chemical fertilizers, such as those that can adjust nitrogen and that has expanded the capability of agricultural production beyond all previous understanding. We overcame the limitations, and we adapted to a new point of view. We explored, discovered and changed the frameworks used to view our world and future.

As we harness more technology and deepen our understanding, we will experience space as an extension of the canvas where we operate. All companies and organizations will benefit from New Space. This is true whether they consume the utility of satellites (in their phones and data networks), technology created for space exploration, or they understand our universe better using what we have learned from our interstellar journey.

Every company will be a space company. Every company will NEED to be a space company.

Introduction

We will do all business and work on a broader canvas that includes space.

We are embarking upon one of the most exciting moments in human history: a time I call "Stellar Singularity." This shift represents the time and age where we see a robust spacefaring economy, a moment where technology, life, and business all operate together across the cosmic landscape.

Many factors are at play guiding us to this time and space, and when it comes to **Stellar Singularity**, we need to travel a little differently. We have to understand what will provide momentum and how we collaborate across our world as we launch farther beyond it.

This book will provide a roadmap to understand the acceleration of increased participation and activity in outer space exploration and the tectonic shift in how we experience progress, work, and collaboration in the new economy above the earth's orbit. The **Stellar Singularity** is the shift as we navigate the convergence of three forces - the acceleration of technology, better understanding of life and sentience, collaboration, and use of space as a Global Commons.

These changes and convergences lead us into outer space and will guide our journey there. We are moving along the same plane in our adventures into the great beyond, toward this new adventure: **Stellar Singularity**.

Introduction

This particular singularity is something we will experience as we journey and operate across vast geographies. The evolution we see unfolding around us will also shape this new landscape of space. This book focuses on helping us reflect upon the fundamental principles and guidelines we need to intentionally embrace as we expand into a new landscape—or spacescape if you will.

ONE

Every Company is a Space Company

NOW IN 1492, Columbus sailed the ocean blue, but he didn't reach his intended destination. He didn't get to India. His initial estimates of the distance traveled were incorrect, and people who knew better pooh-poohed the calculations. But he persisted and did make landfall in what he coined the West Indies, islands that, as it turns out, are nowhere near.

The point is, he got *somewhere*. He found a place and opened the door to new exploration, the New World. It wasn't Columbus who reached what would one day become America. It was the second wave of  explorers. Columbus wasn't seeking a new land. He was supposed to pursue a more efficient route to an established land. That's why America isn't called Columbus. It's named after another Italian explorer, Amerigo Vespucci.

When it comes to space exploration, we've hit that point and started moving past it. We are still determining what we're going to find. What planets will we discover? What will we do when we discover them? Who is calling the shots? Who will make the rules? We are balancing human awe and wonder and human greed if we pay any attention to our track record. There are positives and negatives. There are possibilities both promising and dangerous.

Back then, we had the Old World, the New World, and, of course, empires at play. We still do. Our world hasn't changed that much in 500 years, but it's about to. Commercial space exploration is now past the 1492 point. Early explorers have gone through; there's proof of concept. Now it's starting to take off. People are heading out there. Settlers are looking for their place. We're seeing the Promised Land. The Quakers have left the shores of England.

The Age of Commercial Space Exploration is Here

The age of commercial space exploration is here, and the amount of investment is enormous.[1] Based on available data from mostly the U.S. and Europe, Bank of America expects the space economy to grow at a compound annual growth rate of 10.6%, growing 230% – from about $424 billion in 2019 to about $1.4 trillion in 2030. There's exponential growth, over $450 billion this past year alone. Many different technologies are driving the shift.

Further, China, India, UAE (United Arab Emirates), and KSA (Kingdom of Saudi Arabia) will influence the value of the space economy in ways we cannot fully estimate. So while it's challenging to gauge China's investments, some experts also estimate that

China's space economy alone will exceed two trillion by 2030.

Technology acceleration is enabling and fueling this space journey because we're finding that, along with technological innovations, the pace of computing is also increasing. We're able to do the things we once dreamed of.

The 1920s launched the start of this journey led by the three progenitors of modern rocketry, Hermann Oberth, Robert Goddard, and Konstantin Tsiolkovsky. Oberth was an avid admirer of Jules Verne and, in his teens, envisioned a "recoil rocket" which used the force of expelled gas to launch into space. Years later, he published "Die Rakete zu den Planetenraumen" ("The Rocket into Planetary Space") in 1929, discussing a rocket carrying a telescope off the earth, and launched his first rocket in 1931.[2]

When NASA launched the Hubble telescope in 1990, it opened up new vistas for us. But back then, we were limited by where the Hubble could go and where it could be navigated.[3] [4][5]

Similarly, the Square Kilometer Array Observatory (SKAO) is a state-of-the-art radio telescope with two SKAO telescopes located at radio-quiet sites in South Africa and Australia and auxiliary facilities for the telescopes' operation.[6]

The James Webb Space Telescope (JWST), developed in partnership with the European Space Agency and the Canadian Space Agency, showed us much more. A new lens opened up new worlds.[7]

One of the first things JWST was able to show us was that "The spectrum of light — which contains information about the makeup of a planetary atmosphere 1,150 light-years away — reveals the distinct signature of water" per the European Space Agency. There are indications of water, haze, and clouds in the planet's atmosphere previously unknown on the exoplanet WASP-96b, one of the more than 5,000 planets confirmed to exist in the Milky Way galaxy outside of our solar system to date.[8]

The JWST represents a world of difference in a few decades. A universe of difference. We went much further, all uncrewed, to a point in space where it has reached the destination intact. Not only that, but the JWST takes us beyond what we have imagined. That's its purpose, after all, to open the aperture of knowledge.

Anyone with a computer, television, or smartphone can view new universes of possibilities. So the acceleration of technological capability opens up what we can imagine even further and in broader ways. Fueling acceleration, this time, is the democratization of technology and information.

In the past, only governments had the funds, the technology, and the access to scale and participate in space exploration. Now private individuals—billionaires still at this point—and smaller nations can participate. In my previous book, I talked about the Mangalyaan mission to Mars, wherein India could get to Mars on a low budget, spending 67 million USD, costing less than what it took to make the

Hollywood movie "Gravity"! Its cost was also a fraction of NASA's Mars Maven, which needed 583 million USD.

That democratization of technology is also democratizing access to space. The amount of information we can gather and learn is dramatic - check your social media feed for images from the JWST to illustrate how far we've come in sharing information globally at lightning speed. The ability to have spy satellites, or Global Navigation System Satellites (GNSS), is no longer restricted only to a few large economies or powerful countries. India and Japan have satellite systems now, and other countries may follow them.[9]

Anyone who can pull together the resources to put a satellite into orbit has access to information about floods, crops, and navigation. That includes information they shouldn't have too—information that previously might have been the purview of national defense systems—so there are reasons to be cautious, careful, and aligned in the policies we will form around this new world.

We're in the midst of this expansion of everything we know. We used to think that the atom was the smallest unit of matter. But then we discovered that the atom could be broken down even further. In 1897 J.J. Thompson discovered the electron. Rutherford discovered the proton in 1917 and predicted the neutron, which was discovered experimentally by Chadwick in 1932. Since then, we've also discovered a new world of entities and units of energy we didn't know existed. We now know that twelve types of matter particles make up everything in our universe. These include up quarks, down quarks, neutrinos, taus, muons, and others; all interacting with each other with three forces (leaving gravity, the fourth, weakest, and least understood) and given mass by the recently-discovered Higgs boson.

This rapid acceleration also shows us that we can use technology to understand the quantum movement of electrons, so we can understand what's happening within an atom, within the electrons and protons and neutrons and how quantum mechanics impact them at the sub-molecular level. And harness it all for computing. In the same way, all the changes we see now create the potential for us to know more and do more—and then to know even more and do more again.

On Track to be a Trillion Dollar Industry by 2030

Circling back to the space economy figures we discussed, that significant growth, from $344 billion in 2019 to 2022's $469 billion and projected at $1.4 trillion by 2030, opens up new opportunities that will continue to transform life on our planet. Of course, with those big numbers, investing in space also becomes a driver.[10]

The European Geostationary Navigation Overlay Service's Galileo and the European Earth Observation Programme's Copernicus represent more than €8.2 billion in overall investment between 2008 and 2020.[11] Some calculations show that investment will spur €11.5 billion in cumulated economic value, including creating more than 17,000 jobs.

The Space Foundation—a nonprofit organization, founded in 1983 to support the global space ecosystem—has been tracking the industry's growth for many years. The organization's "Space Report" for the second quarter of 2022 demonstrates the remarkable increase in interest across the globe.[12]

According to that report, the space economy grew to $469 billion, an uptick of 9 percent from 2020.4 Much of the growth, and the money generated, came from the commercial sector, which showed a 6.4 increase in revenues—more than $224 billion from products and services delivered by space firms. Another $38 billion was spent on infrastructure and support for commercial space endeavors.

The overall space economy figure represents the industry's fastest growth since 2014, with something like 9 percent growth year over year. Over 90 percent of the 1,000 spacecraft launched in 2020 were backed by commercial interests—first, the most prominent segment hailed from Starlink satellites from SpaceX. That private investment, plus mergers and acquisitions, is a significant factor in the change we're seeing.

What we need to bear in mind when considering the space sector is that it encompasses many different domains and disciplines. It's a complex area with much ambiguity around policy and regulations. There are huge dependencies between the global communities that participate and nation-states. Given the access and use of space, there are significant implications, too, regarding how spacefaring affects the environment, society, and businesses. There are several factors from a technological perspective that are sparking the kind of significant evolution we've seen in recent years.[13]

We're seeing an emphasis on the commercial value of space over government or military uses. Outer space will soon serve as a medium to provide access to content and services. Ultimately, this leads to the consolidation and integration of technology and services. We're also seeing the democratization of space, with increasing activity from the Middle East, Africa, and Southeast Asia; as people around the globe use mobile data and bandwidth, the need for capacity increases. Also, we expect to see a decrease in the cost and effort to launch space vehicles—India's Mars Orbiter Mission, Mangalyaan, which we touched on earlier, is an excellent example of a more cost-efficient approach—especially with reusable vehicles. Some estimates place this anticipated cost savings at 40 percent.

The Four Value Streams of Space Exploration

The interest and investment have spread organically, and so has the excitement across four value streams:

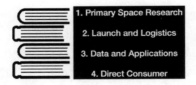

The first value chain or value stream is where we develop technology to explore space. In other words, the technology used to explore space creates other technologies and products. The research and technologies that go into fueling spacefaring work and the challenges unique to space exploration spur more interest and investment in further exploration.

The technology that went into the 1977 Voyager Mission launch provided part of this foundation. Similarly, some of the work around launch systems and microgravity provided challenges—and the solutions we discovered benefit everyone. For example, biology and human health research profit from access to microgravity. Understanding the organism is increased by information on how various species respond to a microgravity environment. We can also study how specific biological functions change in the absence of gravity, as well as with levels of radiation exposure, various mixtures of gases, free electrons, and ions in the thermosphere, all additional features of the space station environment.

Microgravity affects various processes, including crystal formation, fluid mixing, gas-liquid separation, heat transfer, solidification, and combustion. All of this advances study in materials science, with considerably reduced nongravitational noise sources, access to the vacuum conditions of space, extreme heat and cold, enormous distances, and intense pressure for physics study.[14]

Then we have an entire logistical stream of work that has to do with getting vehicles off the ground. Companies like Airbus, Boeing, Northrop Grumman, Blue Origin, and SpaceX operate in that space: the launch or logistic streams. They focus on getting "stuff" out of the range of the earth's gravity and into orbit. In addition to launch, significant work is involved in satellite maintenance - extending satellite life by adding propulsion modules while in orbit, delivering supplies to the International Space Station, and returning materials to earth. We'll also see the addition of new capabilities to satellites while they are in space to repair, inspect, and replace expended fuel.

In addition, there is the data stream with all its applications —all the services we can create and use to benefit life here on earth. Think of satellite services, television, radios, and cell phone towers all as value-added services when we look at them through the lens of our adventures into space (and all the work that goes into them). These companies like Sony, Amazon, and Microsoft have their satellites and use that data to serve clients or companies.

A fourth industry stream is the direct consumer category, composed of governments, individuals, and companies that may use the data we get to make estimates and decisions—or navigate from one place to the next. This category also includes space tourism which will increase with more commercial missions.

The Ever-Expanding Space Economy

The boundaries between what we have traditionally thought of as space and non-space activities are increasingly blurring. Think about something like the telephone. It's not something we needed space for, to begin with, but now we use landlines and wires less and less. Cell phones function without using satellites or space today, but we're already seeing that change and the evolution of technology will continue to provide new pathways. If you take an integrated view of what's happening in different industry sectors, a whole host of technologies developing will change how we do business and live our lives on a day-to-day basis. All can be, or eventually will be, connected to developments for the space economy.[15]

There are different emerging categories of companies in the space industry: launch vehicles; launch assist; materials and components; data analytics; software and ground; satellites; and operators, which include satellite repair, refueling, mining, manufacturing, tourism, communications, earth observation, testing platforms, space debris (the correct term being Resident Space Objects) removal, settlement, and military use.[16]

Launch vehicles include any company that helps get satellites or other instruments into the atmosphere. Think of companies like Virgin, Blue Origin, SpaceX, etc. The second category, launch assist, supports those launches. Then some companies build satellites, and many others work with data and developing software.

You also have companies providing test platforms and companies removing RSOs. This has been a topic of discussion because a small object can cause significant damage, like when a small object traveling at high speed penetrated the Space Shuttles windshield in 2016.

The ESA and NASA believe that something tiny, no more than a few thousandths of a millimeter across, crashed into the window at high speed and caused the damage. Likely, candidates are a paint chip or a small piece of metal. A  company called Off-World currently leads the Air Force Research Laboratory contract to remove RSO.

There are even companies directly focused on settlement, such as Astrobotic Technology and Astroscale Holdings. Household name companies like Amazon and Microsoft are also very active in space. Amazon and the Amazon Web Services (AWS) Ground Station provide satellite antennas to access AWS services. The company, which started as a book retailer, operates fully managed ground stations, onboarding and scheduling satellites, identifying contact windows, and scheduling contact with satellites. It also commands, controls, and downlinks data, receives, processes, and distributes it for more cost-effective storage.

You can also download and process data faster and with greater ease. It also allows reduced processing time for use cases like weather predictions or disaster monitoring. Similarly, since 2020, Microsoft has been working on providing a new service called Azure Orbital, which connects satellites directly to its cloud computing network. They work with companies like ViaSat, Kongsberg Satellite Services, and others.

Everything we do on earth today will eventually have some extension into space—again pulling us closer to the moment of **Stellar Singularity.**

The continuous research and design efforts, with applications for supply chain optimization and so many others, will only increase in the coming years. There's also significant development around ground stations and cybersecurity, which goes hand-in-hand with governments and companies needing to secure data as it's increasingly managed via satellite systems. We can use these technologies for the Internet of Things and manage all the data that supports it.

TWO

Exploring Space Improves Life on Earth

IN MANY WAYS, we aren't looking ahead to an era of space voyages. It's already here. Just take a look around you. The technology developed to support space exploration—and its applications—already permeates our lives. Even technology designed to address secondary needs has become part of our lives, at least on the periphery.

Just think of freeze-dried ice cream you might have had at birthday parties as a child. Freeze-dried ice cream and other freeze-dried food were first created by NASA for astronauts. Now we enjoy it here on earth. Similarly, the material used in everyday lifestyle items, like foam bedding, was created for space utility. NASA has made a business of allowing businesses to leverage technology developed as space tech.

Since 1976, NASA's Technology Transfer Program has shared innovations developed for exploration and discovery with the public. One interesting example: NASA has long worked to mitigate bone loss caused by extended stays in space. This research started with engaging biotechnologists to test osteoporosis treatments on mice in microgravity to help astronauts and the rest of us here on earth. The results supported research that helped shape a new bone treatment drug and others in development. Similarly, NASA worked with an entrepreneur to develop methods to help astronauts retain muscle mass in microgravity... and the Bowflex® Home Gym was born.

Let's move on to the issue of water. Clean water is a concern on earth; it also presents complex challenges for outer space exploration. Every pound of water shipped into space costs thousands of dollars—so sweat, urine, and breath moisture are recycled. You can imagine how those technologies provide solutions to the need for purified water on earth (and the companies who have dived into the markets surrounding that need). From indoor farming techniques to the world's first industrial-strength robotic glove (a partnership between NASA and General Motors), finding solutions for outer space has continued to create pathways to develop technologies that improve our lives on earth.

Outside of hardware, machines, and processes, scientists developed countless software applications for use in space, and those, in turn, can be applied for the betterment of the earthbound.

One recent example: NASA has released its design and integration tools, including a Mixed Reality Engineering Toolkit, and created augmented reality and virtual reality tools for integrating spacecraft designs and real-time mission telemetry for multiple domains over a mission's lifecycle. Currently, this is available only for use by federal employees and contractors, but it will eventually make its way into mainstream commercial development as it has in many other instances. This technology can help businesses reduce cost and errors in concept design and assembly, reduce the time for design and development of assembly and tools without costly 3-D printing or manufacturing, reduce the need for mockups and travel, and reduce costs and time overall. It's also technology that will enhance how we use the Metaverse.

Another vital application of space technology is access to noninvasive observation tools, providing better and more accurate data. We're seeing increased use of Earth Observation and Global Navigation Satellite Systems (GNSS)—these are constellations of satellites providing signals from space—has significantly accelerated how we connect and use information. While GNSS determines a precise position anytime, anywhere on the globe, Earth Observation provides information about the earth's surface, atmosphere, and marine systems. Using both technologies together unleashes exponential potential. It can also offer robust connectivity for areas where land-based connectivity is challenging to deploy, such as deserts, mountain ranges, islands, rainforests, and wetlands.

The European Union Space Programs estimated that 5 billion GNSS devices are in use worldwide, with three general benefits:

- Transportation
- Mass-market use through smartphones and other devices
- Professional services like surveying and mapping activities

These programs provide multiple benefits and cross many disciplines and industries. They enable crop production optimization by blending geospatial data and land humidity measures along with high-precision agricultural techniques. That may result in increased crop yields—by more than 10 percent—plus reduce the demands for fuel, fertilizer, and pesticides by as much as 20 percent.

A few other highlights from these flagship projects of the European Union: significant benefits for developing sustainable smart cities, precise construction surveying and infrastructure monitoring, and improved traffic management.[1]

All those are significantly critical to urban planning and development because more than 60 percent of the world's population is projected to live in cities by 2030. Ultimately, these developments can help our cities become greener and less polluted, plus help predict natural disasters to improve the quality of life for billions of people.

Space Technology for Global Poverty Reduction

These technologies can help us globally with some of our most pressing problems. Take the United Nations' 2030 Agenda for Sustainable Development as a snapshot of challenges and opportunities.[2] With 17 Sustainable Development Goals, it can show us what some world leaders consider critical to address and provides a comprehensive framework. This includes goals like progress on climate action; peace, justice, and strong institutions; affordable and clean energy; and eliminating hunger and poverty—to name a few. Many of them can immediately benefit from the use of space technology.

Take poverty, for starters. We continue to see nations with economically disadvantaged communities where services have yet to move forward to have an infrastructure to support improving conditions for impoverished people.

Technology can help in many viable ways to help change this: It can forecast natural disasters and coordinate better aid by mapping populated areas and access to services. Poverty is a broader problem than most people realize: one in five people in developing regions live with the equivalent of $1.25 per day. Even today, we have millions of children under 5 years dying each year.

Using GNSS and Earth Observation data can help agriculture to be more productive, with documented increases of over 10 percent while reducing pesticides and other interventions.[3] Remote sensing satellites comprise approximately one-third of all operational satellites currently in orbit, and we can measure only over half of the essential climate variables from space. This is incredible progress if we put it to use cooperatively. The government of the Dominican Republic has used space data solutions to eliminate disaster risk for a country long plagued with hurricanes and tropical storms. The destruction of homes and businesses contributes to poverty. We can anticipate much of that loss and the risks mitigated.

The capacity to collect information about adverse weather conditions that might impact not only homes but also crops —or the ability to foresee potential droughts or floods, which we can track using remote sensing satellite data—is a game-changer. It affects housing, food security, and many other factors directly impacting poverty.

Also, expanding energy availability is one of the most effective means for reducing poverty and increasing quality of life. Today, one-quarter of the earth's population has no access to electricity.

Trying to provide energy to developing nations with conventional sources will increase pollution and climate damage and require expensive delivery of fuels. If we beamed down large amounts of energy to terrestrial receiving stations from Earth orbit, it would be pollution-free, greenhouse-gas-free, logistics-free, and radioactive-waste-free. And we would have a 100% availability solution to provide energy to developing nations, especially those near the equator.

Microwaves may be used as vehicles to harvest solar energy from space. Satellites in space can collect more energy due to the lack of atmospheric interference and 24 x 7 collection without interruption by night. Project Cassiopeia consists of a series of strategically placed satellites in a high Earth orbit with the promise of harvesting solar energy and bouncing it back to earth. These satellites could produce all of the world's energy in 2050. Downsides include the expensive cost of launching satellites into space and the carbon dioxide it would create. The host of this project, Space Energy Initiative (SEI), predicts that the timeline for something large scale is heavily dependent on private investors seeing that government funds are currently limited.

Space Technology for Sustainability and Infrastructure

We can also apply space technology to support clean and affordable energy—another of those United Nations Sustainable Development Goals. It can fund infrastructure monitoring, synchronization of the power grid, and more.

Satellites called "power sets" can also collect solar energy and stream it to receiving stations thousands of miles away. This brings us back to the exciting potential of capturing solar energy in space and beaming it back to earth.[4] This idea, as I mentioned, was first proposed in the '60s, propelled by China's efforts, and spurred on by the European Space Agency's moves with a feasibility program called Solaris to find clean energy sources.

China has plans to deploy an operational satellite to harness solar power by 2030, and Japan and India have expressed interest in doing the same. The benefits are relatively obvious: by gathering sunlight in space and converting that to microwaves that we can beam down to earth, we are achieving a low- carbon-emission power source that places no burden on the environment. The cost is high now, as new technologies most often are, but all of these technologies and advancements are converging to make them less cost-prohibitive and more viable.

The cost of launching has already been reduced dramatically, especially with SpaceX models that offer more prominent boosters instead of more costly smaller ones, a move that may help the price per kilogram of payload from $10,000 or $20,000 to $100 per kilogram. We're also now seeing more modular systems with small satellites—cube satellites—which are less expensive and easier to mass produce. And more reuse is happening than ever before.

Tapping into solar power isn't the only opportunity to make our world and atmosphere more sustainable through efforts in space. China just finished approvals for a nuclear reactor that will supply power and propulsion in outer space and is estimated to generate one megawatt of electricity for spacecraft power supply and propulsion.[5] This is about ten times the amount of power the ISS (which operates about 120 kilowatts).[6]

Agriculture creates about 25 percent of all our greenhouse gas emissions here on earth. That's more than transportation, to put it into perspective. The agricultural efforts to sustain the meat industry are the most significant contributor. When you think about how people continue to eat more meat, and we are to grow our global population beyond 10 billion by 2050, we are approaching a crisis point (one that makes me think back to Rev. Malthus and his dire predictions).

Now a company called Air Protein has developed the technology to take carbon dioxide and use it to cultivate hydrogenotrophic microbes and feed them a mix of carbon dioxide, oxygen, minerals, water, and nitrogen to create protein-rich flour. This protein, shaped from the air, has been suggested as a solution for the earth and outer space. It doesn't require any land at all; it uses renewable energy. What's most interesting about their work?

We're also starting to see some countries developing spacecraft to provide nuclear generation in space (a topic the UN covered in 1993 in a document called "Principles Relevant to the Use of Nuclear Power Sources in Outer Space").[7,8] The technologies, and their implications, are only limited by our imagination. And they happen to flow both ways.

3-D printing, while it has the potential to provide profound benefits here on earth, even the capability to turn wood flour into furniture, also is making its way to space. In 2021, NASA used a 3-D printer on the International Space Station to manufacture the first 3-D printed object in length (apart from the printer itself: a faceplate for the extruder printhead), illustrating the ability to apply that technology in microgravity. This is significant because it could reduce the risks and costs of space exploration by providing an avenue to create replacements and a just-in-time, on-demand supply chain for parts and tools in space.[9]

Now back to Earth: The hypersonic aircraft company Hermeus, which works with NASA, has acquired the technology to print portions of a hypersonic airplane. These 3-D printers will help make airplanes that can run at a blazing 3,800 miles per hour. That's Mach 5.

And, of course, developing space technology increases the pace of further space technology development. Changes in hydrogen production are just one example of revving up what we've seen fueling these developments. Hydrogen, one of the critical components of rocket fuel (and other types of power), can now be produced much easier than in the past. Scientists used a relatively simple method incorporating aluminum nanoparticles, which take out the oxygen from water molecules, which leaves hydrogen. This creates a large amount of hydrogen and can be done at room temperature, clearing significant barriers to hydrogen fuel production. This new technique provides the ability to produce hydrogen fuel with any water, including wastewater and ocean water. So a core challenge to creating rocket fuel has been cleared—and if we do discover water on the moon or Mars, we can create or manufacture rocket fuel from there.

Mining the Moon

The moon has long been a subject of our adoration here on earth.

It's been woven into lines of poetry, referenced in religious texts, and used to describe our affection for one another. If someone believes you "hung the moon," that's one way of saying they think you are incredible and astounding. The entire world was captivated when humans took their first steps on our moon, our nearest and most familiar heavenly body.

But the moon has even more to offer, as many commercial organizations and countries have come to discover.[10] Mining on the moon has become a point of particular interest across the globe.[11] Companies like Astrobotic, Planetoid Mines, and Honeybee Robotics are targeting lunar resources to support Artemis Base Camps and Lunar Gateway missions.

Multiple countries signed The Artemis Accord to consider how we return to the moon and establish a long-term lunar base.[12] I recently attended the launch of Artemis I, a U.S.-led program attempting to set up a lunar base. (China and Russia have similar aims in partnership with other countries.) Of course, there are also many commercial ventures in this arena, companies like Astrobotic, Planetoid Mines, and Honeybee Robotics targeting lunar resources to support Artemis Base Camps and Lunar Gateway missions. Everyone is still interested in the moon.

Having a base on the moon would provide a lunar gateway, such as the Artemis Mission's purported goal. It could provide a launchpad to explore outer space. Another goal could be to utilize the resources on the moon. There is a line of thought that, by confining our efforts to mine minerals from the moon, we can protect the earth and pull resources from a different source than the planet where we live.

There's much work to be done to determine what minerals are available, and we are starting to make inroads into technology that can extract some of the minerals that are valuable and useful to us. The most important mineral is still water—needed for life as we know it but also to power rockets fueled by hydrogen. Finding water on the moon and the poles of the moon are an essential driver. Still, it also promises, potentially, neodymium, used to build magnets, and energy production from helium-3, a light, stable isotope of helium.[13] We also learned that China discovered a new lunar mineral in lunar basalt moon samples, a crystal now called Changesite–(Y), a phosphate mineral and columnar crystal.[14]

With many systems and subsystems required to extract and process lunar resources, one company Planetoid Mines is validating a suite of tools for in-situ resource utilization or

ISRU. This process extracts and processes the lunar mixed-minerals on-site to produce lunar-derived composites and further beneficiates the mixed regolith into concentrated refined minerals for additive manufacturing.

These efforts help us crystallize and clarify how we might extract more oxygen from the moon's surface. Calculations by scientist John Grant have said there is enough oxygen in the top layer of the moon, which, if accessed - can allow 8 billion people to breathe for 100,000 years.[15] For context, humans need around 800 grams of oxygen daily.

We know the moon's atmosphere is very thin, so when we talk about the moon having all this oxygen available—enough to sustain life on the moon—what does that mean? How would that become available to us? The idea is that we would extract oxygen from the minerals on the moon: lots of oxides like aluminum oxide, magnesium oxide, ferrous oxide, etc. All those oxides contain oxygen, but it's all trapped in a solid form, not something we can breathe or consume. The regolith, or layer of unconsolidated rock that forms the moon's surface, is estimated to be around 45 percent oxygen. We know the raw material for oxygen is there. If we use techniques like electrolysis to extract it, we'll have sufficient oxygen to be useful. If we were to use solar power to fuel mining, we wouldn't be expending vast amounts of energy.

The Mars Oxygen In-Situ Resource Utilization Experiment (MOXIE), led by the Massachusetts Institute of Technology (MIT), is an experimental device on the Perseverance rover intended to convert

carbon dioxide, which makes up around 96% of the atmosphere on Mars, into breathing oxygen.[16] The machine has operated seven times since February 2021, putting out approximately 0.2 ounces (6 grams) of oxygen every hour, comparable to what small trees produce on earth.

The scientists believe that a bigger version of the device than MOXIE would be able to make breathing oxygen for future humans living or working on the Red Planet. If Martian explorers cannot produce oxygen, bringing supplies from earth would add unnecessary heft to a spacecraft. Additionally, MOXIE's goods could be employed as a component of rocket fuel, which is essential for preventing one-way missions. To launch off Mars, a rocket would require 33 to 50 tonnes (30 to 45 metric tonnes) of liquid oxygen propellant.

Many questions remain: how it happens, which equipment we'd need, and many other factors. We have necessary work to complete that cycle. Some commercial space exploration companies, like the Belgian startup Space Application Services (SAS), are setting up an experimental plant to produce oxygen through electrolysis and plan to send some of this technology to the moon by 2025, working with the European Space Agency.[17][18]

With many systems and subsystems required to extract and process lunar resources, one company Planetoid Mines is validating a suite of tools for in-situ resource utilization or ISRU. This process extracts and processes the lunar mixed-minerals on-site to produce lunar-derived composites and further beneficiates the mixed regolith into concentrated refined minerals for additive manufacturing.

Things are moving quickly, not just in terms of moon mining but in the broader picture of space technologies, and we'd better buckle up.

And so the cycle continues. It touches everything from efforts to develop insurance for spacecraft, which will impact our laws and geopolitics, to biological research and medicine. Even studies of how fetuses develop in the womb, an area of microgravity, can help us understand how cells come together, how we might build regenerative tissue, and how to repair or grow new organs. We can also consider how immune systems change in astronauts in outer space, which has implications here on earth. Operating in microgravity, and studying its effects, may help us create better, healthier, and longer lives.

Similar to those studies of bones and the positive outcomes in medicines to treat osteoporosis, there is a world of understanding that opens up to us when we dive into the technologies that enable space travel and explore the knowledge that emerges from it, in turn.

THREE

Private Enterprise Fueling the Next Space Age

NASA CONTINUES to partner with other commercial companies to get rockets in flight and make missions possible. Think about what we saw recently with the Artemis Mission. NASA scrapped the first launch due to a hydrogen leak and some additional concerns, but the plan had been to use the Space Launch System moon rocket, which has four rocket engines refurbished from the Space Shuttle Program from a dozen years ago.

That decision was made to save money. Against this backdrop, NASA has been partnering with and getting bids from other commercial companies to ship cargo to the International Space Station. Now SpaceX has Falcon 9, with first-stage rockets that we can reuse more than ten times, which have been deployed extensively by NASA. So far, the American taxpayers have put about $30 billion into the Artemis program, even before the first launch.

To keep costs in check, NASA has been buying services from the more competitive commercial market rather than trying to build everything in-house. SpaceX has landed a $2.6 billion contract, and Boeing has sealed a $4.2 billion contract deal for Starliner. NASA has also signed agreements with Blue Origin, Nanoracks, and Northrop Grumman to develop designs for space stations and other commercial destinations in space. An interesting note: NASA decommissions the International Space Station in 2031; it does not plan to build another. Instead, the organization is considering becoming a tenant on a space station, presumably produced by a private company. Axiom Space is building modules that will connect to the ISS. When the ISS retires, Axiom will separate its modules and turn them into an independent space station.

Human exploration is also using private missions, as seen by the success of SpaceX in 2020.

The commercial reach of companies in the space sphere continues to grow exponentially. It is the direction of the future.

AST SpaceMobile, a Texas-based public company, is starting to ship satellites that can provide cellular signals directly to mobile phones—and it's sending those test satellites on a SpaceX Falcon 9. These will hold a 693-square- foot solar array, with solar cells on one side to collect energy to power the antenna on the other side. This would use a space-based network to increase broadband globally across the earth. T-Mobile and SpaceX are also collaborating on a new plan to enhance cell phone connectivity using Starlink, SpaceX's constellation of satellites in low Earth orbit, and T-Mobile's wireless network.[1]

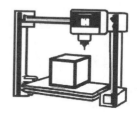

Made In Space, Inc. received a $74 million contract to 3D-print large metal beams in space for use on NASA spacecraft.[2] It's been manufacturing in space, for space" since 2014, when it 3D-printed a wrench onboard the ISS.3 It also launched a commercial version, known as the Additive Manufacturing Facility (AMF), to the station in 2016 and continues to manufacture on the ISS. An example of how manufacturing in microgravity is beneficial is seen in their work with ZBLAN, an optical fiber that, when made on earth, forms crystalline flaws due to the earth's high-gravity environment.

Highlights from Recent Space Explorations

If you're not ready to buckle up yet, I'd like to share a quick overview of the recent space explorations and those that are under development and essential for us to be aware of. I'll keep it high-level here.

We talked about the Artemis mission to launch in 2022 and focus on setting up a lunar base and having a sustained lunar presence for the United States. The other missions are the Indian Mission called Aditya in 2025, which will be putting out a 1.5-ton satellite to develop an understanding of the sun. The JWST Telescope is in place. The Square Kilometre Array (SKA) is being built in South Africa and

Western Australia as the world's largest radio telescope, with hundreds of thousands of radio antennas with a combined collecting area of approximately one million square meters, or one square kilometer.

The UAE sent out Hope 2020, a mission to Mars. Perseverance, which follows Curiosity, launched in 2020 and carried a helicopter drone.

ExoMars, a collaboration between the Europeans and the Russians, will drill two meters into the Martian soil to look for traces of life. The ExoMars project is an ESA astrobiology initiative designed to ascertain whether Mars has ever had life.
One route to doing that uses research into the Red Planet's atmosphere, and the other uses geological studies. The program consists of two missions. The first is the Trace Gas Orbiter, an atmospheric research orbiter launched in 2016 that looks for methane and other trace gases in the Martian atmosphere that could indicate an active biological or geological process. The second - is the delivery of the Rosalind Franklin rover, initially launched by the Russian robotic Mars lander Kazachok (Little Cossack). This mission's initial launch date was for 2020; however, it was postponed to September 2022 because of issues with the lander's parachutes.

Naturally, this never happened because Russia invaded Ukraine in February 2022, and in July, the ESA formally ended its partnership with Roscosmos. The ESA is looking for a non-Russian landing platform to relaunch the project.

The NASA mission of the Double Asteroid Redirection Test (DART) is to study asteroid deflection, with HERA the ESA-led follow-up in 2025. DART was, of course, a resounding success and changed the orbit of the asteroid Dimorphos' orbit by the spacecraft's kinetic impact. Dimorphos orbited Didymos, its bigger parent asteroid, for 11 hours and 55 minutes before DART's collision. Dimorphos' orbit around Didymos changed by 32 minutes since DART's deliberate collision with the planet on September 26, 2022.

We witnessed the first large-scale application of asteroid deflection technology and the first time humanity has purposefully altered the motion of a celestial object.

Juice 2022 and Europa Clipper will study Jupiter's moon for ice and look for traces of life. And then Dragonfly, a drone capable of vertical takeoff under the clouds of Saturn's moon is to be launched in 2026 by NASA, with a similar mission to look for components of primitive life.

The Lunar Gateway was initially considered as part of the Artemis program, but now it's being considered a separate independent program and platform. This Gateway is meant to orbit around the moon. With a polar, near-rectilinear halo orbit inclination, the 40-ton Gateway will be a staging location for later Artemis missions and a meetup point for commercial landers and cargo missions. The same agencies collaborating with the ISS will build the station's modules. There are various modules, including the

Habitation and Logistics Outpost or the Power and Propulsion Element.

Ultimately, the Artemis mission will create the Artemis Basecamp at the lunar south pole. Orion will deliver the first crew to the Lunar Gateway to act as the critical analog for human missions from there to the surface of Mars.

The European Space Agency received authorization and funding to support its planned contributions to the Gateway - International Habitat, providing refueling infrastructure and telecommunications, both of which will dramatically enhance the capabilities of Gateway. Russia also has expressed interest in participating in the Gateway by contributing an airlock.

Gateway will provide a deep space platform to conduct scientific investigations outside the protection of the Earth's Van Allen radiant radiation belts. After the setup of the Artemis Basecamp, NASA will be looking to initiate longer stays on the moon—a month or two at a time. We'll be looking at expanding habitation and other support systems, all to test how astronauts will live on their voyage to and from Mars.

There are many more missions, both commercial and government-driven. All these missions and voyages will accelerate the pace of development and the deployment of space technology, bringing us closer to Stellar Singularity.

FOUR

Understanding Life - Here and Everywhere

YOU DIDN'T THINK you'd be reading about underwater animals when you picked up a book about space exploration.

Yes, we're about to go there.

We can only fully understand space exploration by considering intelligence, life, and sentience. Our intelligence has pushed us to this level—beyond our home planet to discover what lies beyond our atmosphere. And, most certainly, to confront the long-standing question of what lies beyond, including what kind of intelligence and life may exist in outer space. But before we try to look for or explain life elsewhere, we must understand life as we know it here on earth. Life that we continue to learn about and understand; and yet don't know much. We need to learn from what is around us to prepare us for what we are yet to find.

Some of you may have watched the Netflix show "My Octopus Teacher." If nothing else, it will visually depict what we will talk about here. I watched the show and found it fascinating.[1] I also promised myself never to eat another grilled octopus.

The octopus provides an interesting example for us to explore intelligence and sentience. This approach also helps us consider the development of machines and artificial intelligence, which are integral to spacefaring technologies.

Octopuses are cephalopods. They're a little different from other sea creatures and invertebrates because they have different nervous systems and more cognitive complexity.[2] Most of us think of the brain as a nervous system extending its capability to farther reaches of the body. Contrary to this, most of the neurons in an octopus are in its arms. Each arm can independently taste and touch and control basic motions. The other interesting note about the octopus is its giant brain.

I'll explain what I mean: Human beings have about 100 billion neurons, while the body of a regular octopus has about 500 million neurons. This complexity is similar to what dogs and other mammals have. Hence, these deep-sea animals are very distinct from different kinds of invertebrates. An octopus also has a giant brain compared to its overall body size. The octopus genome is also about equivalent in size to ours. It has more protein-coding genes — some 33,000, compared with fewer than 25,000 in Homo sapiens.[3] A few particular gene families are mostly responsible for this surplus. The protocadherins, which control neuronal growth and close-range connections between them, are one of the most extraordinary gene families.

Stellar Singularity

These genes make up 168 in the octopus, which is more than twice as many as in mammals. This is consistent with the creature's egregiously huge brain and the much weirder structure of the organ. Two-thirds of the octopus's half a billion neurons, which are six times as many as those in a mouse, flood out through its arms from its head without the help of long-range fibers like those found in vertebrate spinal cords. *Even after being severed, the limbs' separate neural capability can carry out cognitive activities.*[4] To the extent we can point out unique qualities, the octopus is pretty special.

Moving past intelligence to sentience, let's consider those nine brains—one central to the animal's body and one in each of its eight arms. How might that affect its awareness? Though an invertebrate, the octopus can navigate mazes, unscrew jars, and escape from nearly impossible odds. We've heard a lot about octopuses that run from aquariums. In fact, at the University of Otago in New Zealand, the octopuses figured out how to squirt jets of water at the bulbs, short-circuit the power supply and turn off the lights. We know octopuses also have different relationships with their human keepers, favoring some over others. When you look closely and work to understand their behaviors, you realize that what has been lacking is not the animals' intelligence but our limited observation. Perhaps, it had much more to do with our limited understanding of their capabilities and tendencies.

When Jane Goodall and animal advocates worked against testing on chimpanzees, there was a core belief they battled - that only humans use tools.[5] This rubric meant chimps were; therefore, a lesser species

without the level of sentience
humans possess. As Jane continued her work and broader observation continued, we learned that chimps use tools, and they can even learn languages. Chimpanzees, sharing more than 98% of our DNA, also demonstrate maternal care to territoriality, hunting, and eating meat.[6] Birds also use tools, and anyone living with a dog or cat knows those animals have personalities, preferences, and means of communicating.

Now back to the octopus. It's also true that each cell of an octopus has the innate capability to recognize its surroundings and change its color to match. I'm talking about every single cell in its body. How is that for awareness? And that awareness is not top-down. When an octopus changes color, it's not because a central brain tells the body systems, which communicate to tissues and cells; every cell has that individual capability. Each cell has independent, composable processes without any centralized control - each arm has independent computing power - even when dismembered. Like Kubernetes, the octopus cell can health-check, self-heal with auto-placement, auto-restart, and auto-replicate![7]

In the U.K., a 2021 Amendment to Animal Welfare (Sentience) Bill deemed crabs, octopuses, and lobsters as sentient beings based on their complex central nervous systems. Our rubric for reading intelligence and sentience needs active updating.

The Rosetta Stone of Sentience

When we think about sentience, the most significant factor we have struggled with is the establishment of a rubric. We continue to define and use many frameworks to understand sentience. The truth is that our definition or understanding of what life and sentience means has evolved. For example, if we assume sentience based only on our language, any other language will be incomprehensible. It will be something we need help understanding, absorbing, or seeing. We may still need the decoder ring!

We frequently do not understand or empathize with other living beings.

This is not because the beings are non-sentient but because we may not have the Rosetta Stone to understand what they say. And sometimes, our prejudice makes us want to see them as less.

Our rubric for understanding life is based on what we understand and frequently see on earth. As we encounter different forms of life, such as extremophiles, we must consider how we approach and respect sentience and act as responsible stewards in that environment.

The word "extremophile," for those with a remote familiarity, often evokes images of microbes—simple organisms that can weather any extreme.[8] But extremophiles include many creatures that are not so foreign to us. We can count penguins and polar bears among them. We can also consider different animal and plant life stages as extremophiles. Spores and seeds—and sometimes eggs and larvae—are far more resistant to environmental extremes than mature forms. Elements of extremophile capabilities are also present in animals that can navigate extreme seasonal changes. Many trees, frogs, and insects can make it through remarkably low temperatures over the winter, thanks to seasonal changes in their physiology.

Many have seen illustrations of tardigrades, or "water bears." These can shift into a "tun state," a sort of hibernation that allows them to survive temperatures as low as -250°C or colder in the frigid Himalayas or as blazing as
150°C. They can also survive in the vacuum of space and make it through x-ray exposure. While in this suspended tun state, they can achieve more resistance than the hardy bacterium *Deinococcus radiodurans*—the most radiation-resistant organism known—in terms of surviving extremes of atmospheric pressure. These hibernating water bears can withstand nearly 6,000 times sea-level atmospheric pressure. Tough creatures.

Scientists explore the capabilities of extremophiles by exposing them to high temperatures, U.V. radiation, and salty environments that would be out of range for humans and most creatures within our understanding.

What can these extreme extremophiles tell us about how we try to understand what life represents? Think of the landscape; we are beginning to understand our solar system. Mars presents itself as an inhospitable place. With an average temperature of around -81°F, we consider it a frigid environment. Additionally, the intense exposure to ultraviolet radiation, highly oxidizing, and low atmospheric pressure at its surface make liquid water unstable on Mars. That's not the whole picture, though.

Hydrogeological evidence from the Mars Global Surveyor suggests that liquid water may flow beneath what we can see. Previous evidence hints that there was once a time when water flowed freely on Mars' mysterious surface. It may not always have been a landscape that does not seem to support life or the viability of other types of life. And what does life mean in the first place? This question also relates to the existence and utility of water for life.

Astrobiology, or the study of life throughout the universe, may provide a framework for understanding our beginnings. Astrobiology uses the study of extremophiles because there may be life outside of earth that lives in harsh settings. But astrobiology is much more than just the search for extraterrestrial life. It also looks at fundamental issues such as the beginnings of life, habitats that support life, how life develops, and the boundaries of what life can tolerate.

The quest for the first common ancestor of all living organisms on earth—also known as the Last Universal Common Ancestor (LUCA), the Last Common Ancestor (LCA), or the Cenancestor—is at the heart of astrobiology. But this inquiry leads us to even more fundamental queries, such as, "What is life?" Is all of life simply an ammonia molecule? Similarly, how long ago did the earth transition from a chemical to a biological world? Does life have the capacity to reproduce itself? Something capable of evolving? Extremophiles, those unusual survivors from our history, will undoubtedly play a part in biology's fascinating future as it seeks to answer these origin-related issues.

We can break down living organisms known to us to the molecular level—organic molecules made of long chains of carbon. Most of these require energy from living cells to operate. How did those first organic molecules come together? We've heard talk about a primeval broth, a soup where the first living cells took shape. Underwater hydrothermal vents provide a window into what this might have looked like and how these organic life molecules may have created the miracle of life. Of course, our home planet is not the only place where such broths may be percolating. It's entirely within the realm of possibility, especially considering the vastness of space, that these hydrothermal vents may exist on other planets and moons. When we think of life beyond earth, we visualize a planet like ours because it's what we know. We have experienced the benefits of a thick atmosphere to keep a blanket of oxygen in our lungs. We, humans, enjoy and need a particular and tightly bound range of temperatures.

Some species can survive in situations with no oxygen at all. Extremophiles have "extremozymes," or extreme enzymes, that allow them to survive in hostile conditions. It's crucial to remember that these species are only 'extreme' from a human standpoint.

The concept of panspermia holds that simple living forms could cross planetary boundaries and endure the voyage. Although the idea is sometimes mocked as illogical and speculative, recent research has given panspermia more validity. According to some, panspermia could have been the process by which germs from other planets first arrived on earth and served as the ancestors of all later-evolving species.

Previously thought only to contain three, we discovered meteorites holding adenine, guanine, cytosine, thymine, and uracil.[9] These nucleobases are the five components that make up genetic life on earth. Extraction has proven that meteoritic  samples contain an abundance of these bases.[10] Isomers found in the meteoritic samples but not the soil source of the samples support the theory of an interplanetary origin of life. Based on scientific research, testing the source—pristine asteroids—is the next step in confirming this discovery.

Planetary Protection is the field that enables the responsible exploration of the solar system while protecting both the scientific integrity of our observations and the safety of our planet. By carefully controlling the potential contamination of other worlds and preventing the return of potentially harmful materials to earth, we can better understand the origins of life in the universe and the potential for life elsewhere in the solar system.

NASA's Office of Planetary Protection uses a variety of techniques and methods. These can include using sterilization techniques to remove potential contaminants from spacecraft, developing specialized containment protocols for handling extraterrestrial samples, and creating strict guidelines for spacecraft operations to minimize the potential for contamination of other worlds.

The Office of Planetary Protection also works closely with other organizations and agencies within the United States and internationally to ensure that planetary protection efforts are coordinated and effective. This agency partners with the European Space Agency and the Japanese Space Agency, as well as international organizations such as the Committee on Space Research (COSPAR) and the International Council for Science (ICSU).

By studying the conditions that allow for life, we are working to understand what life may look like elsewhere. It enables us to hypothesize the possibility for life to evolve and survive in environments we would consider wildly different from those we know. And where you and I may not be able to survive. Thermophiles and hyperthermophiles can survive in extremely hot water—think hot springs or hydrothermal vents on the ocean floor where temperatures reach as high as 230°F. *Sulfolobus shibatae*, in acidic hot springs, is a thermophile as well as an acidophile, thriving in extremes of heat and acid. The polar opposite of the thermophile is the psychrophile, which can manage temperatures below the freezing point of water just fine. They love permafrost, polar ice, and frigid ocean water and have developed proteins that work like antifreeze.

Similarly, halophiles use proteins to maintain proper water balance in salty environments that would kill other microorganisms. Those creatures thrive in cold, hot, wet, dry, alkaline environments with an abundance of salt. Barophiles can go where other microorganisms can't because they can withstand high pressure—some have even evolved to be so accustomed to this environment that they cannot survive outside of it.

Oxygen isn't the only material organisms can "breathe;" others can use metal in the same way our bodies apply oxygen's life-giving (to us) power. Shifting our perspective, we could even consider ourselves an example of an extremophile when we consider creatures that cannot live in an oxygen-rich environment. I could go on, but you get the picture: the wonder of life and its resilience extends beyond the bounds of the specifications required by humans.

Moving beyond the science lab, where else can we find guidance for a new lens to understand life and sentience? Consider the ancient religion of Jainism, which we can trace back to the 6th century B.C.[11] It provides 10 "life principles" that can help us take another view. These principles are linked to the five senses: energy, respiration, life duration, speech, and the mind. The system of organization in this religious approach classifies beings according to their sensory capabilities, indriya, life essentials, or prana.[12]

Here's a look at how this is laid out within Jain texts:

> "The one-sensed beings have four—sense organs of touch, strength of body or energy, respiration, and life duration. The two-sensed beings have six, namely the sense of taste and the organ of speech, in addition to the former four. The three-sensed beings have seven with the addition of the sense of smell. The four-sensed beings have eight with the addition of the sense of sight. The five-sensed beings without minds have nine life principles with the addition of the sense of hearing. Those endowed with the mind are said to have ten vitalities with the addition of the mind."

Jivas, or lives, show "Paryapti" (completeness), where they take in matter like food and convert it into separate kinds of energy. There are six kinds of Paryaptis: food, body, senses, respiration, speech, and mind.

In Jain thought, four stages of perception—observation, will to recognize, determination, and impression—lead to subjective cognition, the first of five kinds of knowledge (jnana). The second kind, shrutajnana, derives from the scriptures and general information. Both are mediated cognition based on external conditions perceived by the senses. In addition, there are three kinds of immediate knowledge—super sensory perception, reading the thoughts of others, and omniscience.[13]

This approach gives us another frame for we might consider life outside the one we experience through our experiences (or as we seek to define sentience in artificial intelligence).

Life encompasses far more than what we can take in from the window of our minds, what we know today, the senses through which we experience the world, and even paradigms of how we think.

FIVE

Singularity and the Spacefaring Economy

AS WE CONSIDER how things are lining up to take us farther into space and to improve and transform life on earth, I want to bring us back to ***Stellar Singularity***. The phrase "singularity" derives from the field of mathematics. It describes a point where we are incapable of deciphering exact properties. Let's break that down a little further.

When the formulae and equations don't work anymore, singularity represents the point at which we cannot fully define or create boundaries in mathematical terms.

Ultimately, the best definition of a singularity is an instance in which we cannot predict what happens beyond it.

It is important to note that these estimates are always a function of what laws of nature and physics we understand. The impact of a singularity and its very nature may change as we expand our knowledge.

Singularity also presents a phase we can think of as a watershed point, illustrating a significant change. What does that mean? It marks a difference between what we knew, what happened before, and what happens as we move forward. Imagine we could time travel and go back to 1492 when Columbus went to search for the Americas —or even before that time, to the Magna Carta. If we tried to explain how the internet works to people we encounter, they would be baffled. Imagine describing how an airplane works, or even a camera or Netflix. You'd stir up extreme confusion at best and terror at worst. Or they'd probably think you were out of your mind.

They would have no concept, context, or capacity to fathom the world we inhabit. It would be impossible to describe our current reality in terms they understand. This is what I mean by a singularity of time. At any point when a civilization changes dramatically to the end of a paradigm shift, its rules and technologies take on new formats, and the definitions of the past are entirely inadequate. Those new technologies would be incomprehensible to people who have yet to experience them.

There is a whole body of work and centuries of thought about singularities, so we will touch upon it lightly here. To try and understand a singularity, though, let's go back to my favorite physicist James Clerk Maxwell, a Scottish scientist of the 19th century.[1] He was the first person to talk about singularity in more general terms. Now Maxwell is also fascinating because he had a wide range of interests, was exceptionally far-thinking and laid the foundation for many of the principles we use. Like many other geniuses, he was misunderstood and beaten by his tutors and labeled slow and wayward at age ten. Many years later, Einstein described Maxwell's work as the most profound and fruitful since the time of Newton. Einstein would share the glowing compliment: "I stand on the shoulders of Maxwell."

Maxwell worked on a multitude of things, like developing the Maxwell-Boltzmann distribution, which describes the kinetic theory of gases.[2] I also talked about him in my previous book, *[Block]Chain Reaction: The Future of How We Live and Work*, where we covered "Maxwell's Demon," the proxy for the forces manipulating changes at the molecular or subatomic level.

Maxwell laid the foundation for the line of thinking that led to quantum computing. This scientist, ahead of his time, was the first to use the term "singularity" in a more general sense outside of mathematics, a reference to how small changes can lead to significant effects. He talked about the idea of singularities to argue against absolute causality. Something we should remember when we hear the naysayers and their ominous predictions that we are hurtling toward a world beyond our control, with the imminent take-over by Artificial Intelligence.

Maxwell was also interested in problems of a cosmic nature. When he was a professor at Marischal in Aberdeen, he worked on a problem that people have been noodling on for centuries concerning the nature of Saturn's rings. Quick backstory: Nobody understood how these rings stayed in place and didn't crash into Saturn or break up and move away. In 1857, it was chosen as the topic for the Adams Prize. Maxwell studied the problem and proved that a fluid ring could not exist as it would be forced by wave action to break up into blobs, while a regular solid ring could not be stable. Because neither was possible, he concluded that the rings must be composed of small particles, which he called "brickbats." Fast-forward more than a century to the Voyager I and Voyager II in the 1980s, and we discover that he was correct. Saturn's rings are, indeed, composed of particles.

One good way to wrap our heads around a more uplifting understanding of singularity is to think of what happens when a pane of glass shatters. Everything that happens after is difficult to define or describe because so many forces and variables are at play. This example shows how a singularity is *unstable* because what happens may be disproportionate to the scale of the original cause, or it may have no discernible impact.

Think about the glass in the windshield of a car - when it encounters a single flying pebble. You may be able to drive around with a cracked windshield—or it could cause a five-car pileup on a freeway. It also matters that the glass is of the windscreen - a different outcome and implications if it were the glass of water. So, a singularity is *related to the system or context* around it, and the results depend on that context. Singularities are also *unique* because of the scale; the outcome is not necessarily related to how large the pebble or the windshield crack is. And, since there is not always causality or precedence to define it, it makes it *qualitative*. It's also *irreversible* because we cannot mend the windshield with the same force that caused the break in the first place.

This singularity is also *subjective*. The perception of its outcomes or potential outcomes depends entirely on how the driver or car owner perceives it. The fact that the windshield cracked could have been considered catastrophic or irrelevant, depending upon the mindset of the car owner. The singularity is *random* because we cannot always predict when a pebble may be thrown or fling in our direction from another side of the street. It could have been caused by a child who threw the pebble in fun or malice. Or it could have been flung up accidentally from the gravel on the side of the road.

The variables add complexity, and, notably, they happen when there is an interaction or overlap between two systems, the system of roadways and automobiles, with the system of recalcitrant children.

Now there are several different singularities. One is the mechanical singularity, the point beyond which we cannot predict the subsequent behavior of a machine. There is also gravitational singularity when time and space break, a point we cannot determine the "where" or "when." And the laws of physics, as we understand them, can't explain things. The Penrose Hawking singularity theorems tried to answer the quandaries of when gravitation produces singularities. Black Holes represent singularities.[3]

It's worth considering for a moment the distinct relationship between humans and technology in the book, "The Singularity is Near," by Ray Kurzweil, an outstanding futurist.[4] Ray's work is required reading for anyone wanting to understand more about technology acceleration. He talks at length about the acceleration of artificial intelligence, popularly known as "the singularity" or "technological singularity." His point is that, with human genomics and nanotechnology, we will reach a point where we can alter and augment ourselves at the molecular level, ultimately changing the boundaries and definitions of life and what we understand as sentience.

The science-fiction writer Vernor Vinge popularized the phrase "singularity." In 1993 he wrote about singularity representing a point at which we achieve superhuman intelligence.[5] Vernor said it could happen around 2030, but many experts predict that we are looking at something more like 2045. The fundamental disagreement is not *whether* it will happen but *when*.

All these discussions on singularity focus on the technological singularity or the approaching superintelligence. The point at which machines, or any other intelligence, advances past what human brains can offer. Most often, this conversation on technological singularity revolves around artificial intelligence (AI), a tool we already use in so many ways:

- Translating languages.
- Interacting with chatbots to solve a customer service issue.
- Analyzing massive amounts of data, such as all the Earth Observation data we discussed.

Fear can creep in, but the limits are only in our thinking, knowledge, and understanding of managing and governing these developments.

AI is a core element of how we use social media platforms. We can also use it to diagnose cancer by scrolling through large amounts of data.

We all remember when a computer won out in Jeopardy or when the computer Deep Blue outmaneuvered Russian chess grandmaster Garry Kasparov. All this ability of  neural networks and algorithms to do more with data has life-changing benefits now within our reach. And it's only accelerating.

If we don't construct AI with the correct or sufficiently comprehensive sets of data and rules, it can make some big mistakes. We've seen examples with facial recognition: some AI cannot read faces with dark complexions. Or, we've seen instances where AI systems predict things inaccurately because they need to use complete datasets or consider all variables. Of course, we've also seen negative aspects of artificial intelligence, and that's because artificial intelligence is only as good as the training built into its algorithms. As its capabilities are accelerating, the potential for misuse and poor construction are also accelerating. And as it becomes more ubiquitous, some people need to consider all the considerations for building ethical AI.

The other thing critical for us to understand is the distinctions between what AI can and can't do. The more we are removed from understanding how the technology works, the more at risk we will be. We cannot just be consumers of AI; it will become essential to know how it works. The fewer people who understand the technology, the easier it will be to manipulate datasets and use AI incorrectly. We must commit to understanding black box models to outsmart and override users and operators. Understanding how an AI model derives a decision point is core to this capability. It's not okay to be satisfied with a black box model of AI where we don't understand what goes on inside and how it works.

Similarly, as organizations are collecting data, it's essential to do multiple things:

1. Build in reviews of the data
2. Review the standards of privacy
3. Ensure that the data collected and used to train the AI is the right set
4. Manage data sets appropriately.

There have already been several issues with the use of AI in hiring. That's a whole conversation by itself. There are also other instances of bias in facial recognition, so there are legitimate concerns about using it to make decisions about loans, mortgages, or credit ratings.

Thankfully, the principles for AI ethics have received much attention. In December 2018, the European High-Level Expert Group on AI published the first draft according to which trustworthy AI should be Ethical, Lawful, and robust. They then set out seven criteria for "trustworthy

AI".[6] We need some similar principles to guide our thinking - which I'll cover in the chapter on the Interstellar Magna Carta.

As you know, we have used more and more technology to understand what's within our solar system, within our heliosphere. Some of that began with previous generations and continues through today. NASA sent up Voyager I and Voyager II on interstellar missions in 1977, primarily focused on exploring Jupiter and Saturn. NASA extended the mission after a series of discoveries there, including active volcanoes on Jupiter's moon Io and intricate details of Saturn's rings. By 1989, Voyager 2 sailed past Jupiter and Saturn and crossed the orbits of Uranus and Neptune.

Voyager 1 and 2 have been traveling for more than 40 years. They are now significantly further away from earth and the sun than Pluto.[7] Voyager 1 made the first-ever entry into interstellar space in August 2012, an area between stars packed with debris created when surrounding stars died millions of years ago. On November 5, 2018, Voyager 2 entered interstellar space, where researchers hope to gather further information. The current mission of the explorers, the Voyager Interstellar Mission (VIM), will investigate the farthest reaches of the sun's sphere of influence. And beyond.[8]

Through the Deep Space Network or DSN, both spacecraft are still sending scientific data about their surroundings.

In 300 years, it will reach the beginning of the Oort Cloud, a collection of icy material where many comets emerge. This is likely the furthest material still orbiting around our sun. In 30,000 years, it will reach the end of the Oort Cloud. There's much beyond what we know, understand, and can even comprehend. We'll only get there by working together.

How can we build guidelines for operating in a space economy? We must do so, especially in light of how quickly the ground beneath our feet moves. We can take a cue from the principles laid out in the Belmont Report - Beneficence, Non-maleficence, Autonomy, and Justice to move forward with care and a concern for equity and the common good.[9]

Deep Blue beat Chessmaster Garry Kasparov in 1997. At that time, Deep Blue was a hulking computer, the only one who could play chess at that level. Now the AI that moves digital chess pieces around has improved so much that AI chess opponents are unbeatable. This has heralded a kind of chess that Garry created, an opportunity for humans and algorithms to collaborate. Together, they can create better outcomes. The computer provides computations and shares probabilities with the human being, who then uses judgment and human sentience to make the best choice.

This principle is very aligned with the principles of augmented intelligence. With the rise of AI, quantum computing, and commercial space exploration, the true scope of Stellar Singularity is when we harness all these advances and operate in a different, vaster geography of application.

To that extent, ***Stellar Singularity*** is where we should be if we want to look ahead. We have to look at the context of where singularity is going to play out.

Stellar singularity is where these new capabilities converge, the stage for the most profound changes in and beyond our world.

SIX

Collaboration and Competition in Space

IN 1967, legendary saxophonist John Coltrane recorded the studio album "Interstellar Space," which ended up being a duet with jazz drummer Rashied Ali. The story goes that when Coltrane arrived at the studio, nobody else was there. It was just Rashied who asked whether anyone else was joining. John looked around and replied: "No, it's just you and me ."The story goes that Rashied asked, "What are we playing? Is it fast? Is it slow?" And Coltrane said, "Whatever you want it to be."

John would die months later at just 40 years old. We can only speculate about what was on his mind as he entered that creative collaboration. If the works produced are any indication, the collaboration created something more significant than the planes he had explored before.

Stellar Singularity

Coltrane played a sonic interpretation of what we call "interstellar space." The album has many compositions named after planets, but the extemporaneous composition was music about the space between the stars. Coltrane included elements from the music and philosophy of India into jazz and a much more robust and profound spiritual dimension.

Interstellar space is not empty; it's the realm beyond the heliopause - the boundary within which solar wind flows, and the sun influences the flow of material and magnetic fields. All of which impacts other cosmic bodies. This gravitational pull and magnetism create the heliosphere. Interstellar space is outside the heliosphere—in the regions between stars, and affected by the stars, in the galaxy. Cosmic objects, like Coltrane and Rashied, move fast or slow depending on what they encounter. That rate and rhythm are analogous to the pace of our efforts to explore past our home planet.

Interstellar space represents the final frontier and where we seek to travel as space exploration accelerates. We will leave the protective bubble of our sun into the "almost- vacuum" occupied by hydrogen and helium clouds, along with heavier gasses.[1] Only two spacecraft have made it to interstellar space, though they have yet to go beyond the Oort Cloud. Voyager 1 was the first, after traveling thirty-five years, and Voyager 2 after forty-one years (both taking a longer route past other destinations). Per NASA, traveling interstellar distances would need exotic technology, such as suspended animation or multi-generational life support and solar sails.[2] Getting there is a long time away.

Now Coltrane and Rashied made music that Rolling Stone called "plainly astounding." But their approach of "whatever you want it to be", which is what's happening with the Outer Space accords, is not necessarily the right recipe for success. In the words of Leon Alkalai, an engineering fellow at NASA's Jet Propulsion Laboratory in Pasadena, California, and co-author of a report on exploring interstellar space. "Like the ancient mariners, we want to start creating a map."[3]

With the continued activity in outer space and ongoing conversation among countries and various players in this new exploration, the old rules of engagement don't cover what we need.

We have two crucial treaties (both developed during the Cold War Era) that provide a template: the United Nations' Outer Space Treaty, which, as suggested by U.S. President Dwight Eisenhower, was modeled on the concepts built into the second treaty, the Antarctic Treaty.

The Antarctic Treaty is one generally considered to be a success as it's been effective at getting different nation-states to work together against competing claims. It's served its purpose of protecting a shared resource very well as countries like Norway, Chile, the United States, and others have tried to claim parts of the treaty. It has also dealt effectively with the conversation around weaponization: nuclear testing in the Antarctic. And it has preserved the territory for peaceful research.

The Space Race between the U.S. and Russia began during the Cold War. When the Soviet Union launched Sputnik in 1957, the United Nations established the Committee on the Peaceful Uses of Outer Space (COPUOS).[4] Since then, the United Nations has remained active in trying to formulate policy around outer space endeavors. As of 2016, COPUOS had 77 members and had formed additional subcommittees with secretariat services provided by the United Nations Office for Outer Space Affairs. COPUOS drove the development and establishment of five important agreements that provide an umbrella framework for what we see happening today.

The UN created five principles in the sixties. The first was the Declaration of Legal Principles, which provided the basis for the Outer Space Treaty of 1967.[5] Then the Broadcasting Principles of 1982 related to how televisions broadcast signals without interfering with the signals of other countries. The Remote Sensing Principles are related to the use of electromagnetic waves to collect data through observation and remote sensing. The Nuclear Power Source Principle addressed the impact of radiation on humans for spacecraft using nuclear power. Finally, the Benefits Declaration of 1996 stated that humankind should carry out space exploration for the benefit of everybody. The UN forged that declaration two years before the International Space Station was established.

If there is an emergency landing or an accident, any country is supposed to provide help immediately. In 1968, The UN formed the Rescue Agreement to ensure that astronauts are always supported and cared for, regardless of their nation of origin. Then came the Liability Convention of 1972, the Registration Convention of 1975, and the Moon Agreement of 1979. The UN set these five key treaties in place in the 1960s and 1970s.

A few notes on these additional agreements. The Liability Convention agreement lays out the impact of what happens if something from space damages something on earth. We saw this play out recently with SpaceX debris falling in farmers' fields in Dalgety, a town in rural Australia. Per the Liability Convention, the state where the space object was launched is responsible for damages on the earth's surface. Because the SpaceX launch was in the United States, NASA would be responsible for the damage overseas.

The United Nations Register of Objects Launched into Outer Space provides a mechanism to register and track everything launched into outer space. That's vitally important, increasingly so now, as more and more objects are launched into space. The Moon Agreement provides additional details on the usage of the moon and other celestial bodies. But it has only been signed by 16 countries, which are relatively minor players in space exploration.

With the backdrop of all this technology, information, and knowledge expansion, we still operate on a global plane here on earth. Think of shifts in our geopolitical environment and dynamics. During the Cold War, the U.S. and Russia fought the Space Race. In the post-Cold War era, we've seen many ongoing shifts. China is becoming very powerful. Europe is destabilizing. Russia continues to attempt to flex dominance.

China, in particular, is working to match or exceed the capabilities of the United States in space to gain the military, economic, and prestige benefits associated with space leadership. China's space station began assembly and crewed missions in 2021, with full operational capability expected between 2022 and 2024. China  also plans additional lunar exploration missions and intends to establish a robotic research station on the moon, followed by an intermittently crewed lunar base.

The Chinese People's Liberation Army (PLA) will continue to integrate space services such as satellite surveillance, positioning, navigation, and timing into its weapons and command and control systems, aiming to erode the information advantage of the U.S. military. Counterspace operations will be an essential part of potential military campaigns by the PLA, and China has developed counterspace weapons to target U.S. and allied satellites. The PLA is currently fielding new ground- and space-based antisatellite (ASAT) weapons, both destructive and non-destructive.

In late November, China sent a new team of astronauts to its Tiangong space station, marking the country's first in-orbit crew handover and potentially the beginning of continuous occupancy in space. This achievement is a significant milestone as China advances in its ambition to catch up with and overtake the United States and every other country to become the dominant power in space.

China, which built this space station in less than two years, also launched more satellites last year than any other nation and developed its own navigation system, Beidou. Earlier this year, Beijing released a White Paper outlining its plans for space science, exploration, technology, and propulsion over the coming five years, stating that the country's space industry "serves the overall national strategy." The paper outlines a broad array of priorities for spaceflight, including upgrading and expanding launch vehicles, building satellite constellations, operating the Tiangong space station, planning crewed lunar landings, and exploring the moon, Mars, and beyond. They also plan to collect near-Earth asteroid samples and conduct two lunar polar exploration missions by 2025, launch a Mars sample-return mission, send an uncrewed probe to Jupiter, and land astronauts on the moon by 2030.

The plans include developing reusable carrier rockets by 2035 and establishing an initially robotic and, later, intermittently crewed research base on the moon by 2036. They are also working on a Mars base by 2045, leveraging China's plans to build a nuclear-powered space shuttle by 2040.

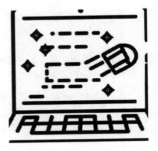

China's space program aims to achieve resource utilization, including asteroid mining, lunar resource extraction, nuclear fusion, and reusable rockets. By doing so, China hopes to establish a strategic presence at the Earth-Moon Lagrange points, positions in space where gravity and centrifugal force balance each other. These plans demonstrate China's ambitious goals for space exploration and its intention to become a significant space power.

In the coming months, Beijing plans to launch the Xuntian space telescope, which is reported to have a field of view 300 to 350 times that of NASA's Hubble Space Telescope.

The political ground beneath us is transforming in ways we have yet to realize. There will be many different players and participants as we explore space beyond our home planet. It would serve us well to be conscious of the ways the geopolitical order is shifting away from what we are used to experiencing and what we might expect. This approach has been attempted but needs to be updated. A few years before NASA launched the Voyager program, the United Nations formulated and proposed the Outer Space Treaty, which we have covered.[6] It was accepted by the world powers interested in space exploration in 1963. Still, it wasn't open for signature between Russia, the U.S., and the United Kingdom until 1967.

This treaty did ratify some principles. Article I states that "the exploration and use of outer space, including the moon and other celestial bodies, shall be carried out for the benefit and in the interests of all countries, irrespective of their degree of economic or scientific development, and shall be the province of all mankind." Article II drills down a little more: "Outer space, including the moon and other celestial bodies, is not subject to national appropriation by claim of sovereignty, by means of use or occupation, or by any other means."

Both these principles are important to remember, but there were also other principles pertaining to space exploration and leveraging it for the benefit of all countries, making a point to note that this realm of territory, if you will, is no specific nation's territory. Other principles are more practical and continue to help us navigate operational issues, such as the concept that countries and states are responsible for national space activities and their impacts, regardless of whether a government or nongovernment entities carry it out. Countries are responsible for damage or contamination. We must use the moon and other celestial bodies exclusively for peaceful purposes—a blanket prohibition on using outer space to extend wartime activities here on earth.

The trouble is that all of this was laid out in 1967 before we started the bulk of our recent space activity. Commercial space exploration still needed to take off; governments conducted all activity. This framework was sufficient for its day.

As more and more countries started making a bid to land on the moon, there arose the question of how we govern all of it. In 1984, the United Nations forged another proposal: the Moon Treaty.[7] This was built around a similar structure as the Outer Space Treaty but focused on laying out the operational aspects of how countries or companies explore our moon. Enough countries agreed to ratify the treaty, so the United Nations adopted it.

Interestingly, none of the spacefaring countries agreed to accept these terms. And it was for good reason. Those two articles of the Outer Space Treaty played a role. So did the fact that the Moon Treaty was based on the United Nations' Convention on the Law of the Sea, which says that the deep sea (outside of territory that any country owns) is the common heritage of humankind.

As a result of that law, an International Seabed Authority exists to regulate activities in the deep sea and charges a royalty, similar to profit-sharing, that all the countries on earth receive some benefit of mineral extraction from the deep sea.[8] Because the Moon Treaty uses this common heritage of humankind language, drawn from the Law of the Sea, the Moon Treaty would imply that you have to pay some royalty to be redistributed among all the countries in the world with any extraction of resources from outer space.

The 2020 Artemis Accords

In October 2020, eight nations came together and signed The Artemis Accord brokered by NASA: Australia, Canada, Italy, Japan, Luxembourg, United Arab Emirates, The United Kingdom and the United States of America.[9] And where did they sign it? In the tiny nation of Luxembourg. Why would such a sprawling undertaking take place there? Because Luxembourg has taken a great interest in operating in and exploring outer space. These new ventures will continue to break boundaries and upend what we are accustomed to seeing.

Notably absent from the Artemis Accords were Russia and China. Russia's view of the Artemis Accords was that they were formulated to support the interests of the U.S., and they did not participate or say much about it. It has been challenging for the United States to cooperate with China on space exploration. In response to the Accords, on March 9, 2021, the Chinese National Space Administration, CNSA, and the Russian Space Agency, ROSCOSMOS, signed a memorandum of understanding to create a joint autonomous lunar base.

The impetus for getting all of this on paper was NASA's intent to build a lunar base, a gateway on the moon. This base would provide a greater ability to explore outer space because it's cheaper and faster to launch from the moon versus planet earth. That's the backdrop. What's interesting about the Artemis Accords, though, is that since the late '60s, we've been trying to formulate some shared perspective on space exploration, how it should be accomplished, and who should benefit. The Artemis Accords were one of the first recent efforts to advance the principles laid out 40 and 50 years ago. It's at least a first step in agreeing that we need *some kind of* regulatory approach. We need to establish additional agreements that detail how we move forward. What our progress to this point does not yet cover are four critical arenas:

1. Signatories - who are the participants, agencies, countries, and private enterprises that need to be part of this conversation? We must include many more relevant players to make the Artemis Accords meaningful and sustainable.
2. Scope - we are headed towards a future that needs to account for more than the moon and a lunar base. Technology is taking us, far and away, past the outposts of the past. So many other considerations will come into play as we explore other celestial bodies.
3. Property Rights - Clarity on property ownership is critical to encourage investment in economic activity. There are no concrete principles or frameworks to enable discussions on how we will manage property rights in New Space. The Artemis Accords haven't touched upon this area of increasing urgency, even as NASA says they

will have humans living on the moon "this decade ."The Institute of Air Space Law in the Hague has established a working group to establish a framework for discussing property rights on celestial bodies. Still, there is a lot of work to be done and resolved.
4. Operational Guidelines. While the Artemis Accords are a step forward, we must consider how much else is required. What are the operational guidelines and regulations we need to put in place? How can we have greater cooperation and coordination in exploring?

The Artemis Accords were set up to allow additional agreements to be added as building blocks. However, it's essential to realize that much more work is to be done immediately to address this new landscape of commercial space exploration. There are so many new players, so starting with eight signatories does not acknowledge that the entire field is different than it has been in the past. Many players are considering space for various reasons that have yet to be explored. Other nations are now looking for ways to extract resources from asteroids and the moon. How we approach the utilization of outer space's resources is another factor that the Artemis Accords need to touch on or with adequate depth. How do we manage conflict?

The emerging landscape needs so much more, focused on addressing much more - for example, tourism, debris, and observation of other celestial bodies and planets.

How do we operate in ways that are fair to all humankind?

A Case Study: Asteroid Mining

Let's take the topic of asteroid mining, which is an area we've seen many emerging concerns around, as an example. Look at the sampling of the asteroid Bennu, which has stirred interest and discussion among many countries. We are on the brink of a rush toward new resources and must contemplate it in the broader public interest.[10]

Are there any efforts to do something in this space safely and conflict-free? How might we extract metals and resources from the moon or other asteroids? Whose jurisdiction will govern those operations? What protects a private company or nation from not being challenged in its efforts? Alternatively, who is watching over how the activities of that private company might affect the rest of us? Which authority provides those rights or implements penalties when things go wrong? Who will license them to allow for those operations? And how do we do all this safely, so it doesn't impact other space objects or satellites?

There are no clear guidelines and no path to how we create them. Asteroid mining is just one illustration. For example, it highlights all the different risks and roadblocks that need to be removed to access resources that might benefit or protect us from practices that could harm us. In this context, the Artemis Accords highlight that the United States will interpret the Outer Space Treaty in a way that allows space resource extraction with safety zones to avoid harmful interference and ultimately establish exclusive economic zones. We need some agreements and laws to cover operations on the moon. Without this, there will be continued confusion and lack of confidence to undertake lawful commercial activities.

To address this, the Space Treaty Project has drafted a Model Implementation Agreement. That agreement defines the "use of resources" to include any location on the moon for any purpose. Any use will be supported if the private activity is authorized or supervised by a country that has adopted the Moon Treaty and the implementation agreement. The current Model Agreement is being discussed, reviewed, and considered a possible alternative to the Artemis Accords. This is important because it would create one enforceable international agreement—terms that are binding and comprehensive for private activity—and an overall framework for international cooperation, including controlling laws and avenues for dispute resolution. None of that is currently included in the Artemis Accords. We are racing ahead, fast and slow. Right now, despite the engines revving in the background, what's holding us back from Stellar Singularity is agreements and treaties that are not fully equipped to help us navigate into the future.

New Rules of Engagement Based on The Antarctic Accord

One model that could serve us well is the Antarctic Accord.[11] Antarctica is also a space beyond immediate reach and beyond the purview of the native human population. Its history may provide insight into how we might consider the moon and the farther reaches beyond our atmosphere.

The Antarctic Treaty Secretariat governs the Antarctic Treaty System—the main treaty and other related agreements and then a secretariat that administers and implements the treaty system.

The main treaty was signed in 1969, with 12 original signatories (Argentina, Australia, Belgium, Chile, the French Republic, Japan, New Zealand, Norway, the Union of South Africa, The Union of Soviet Socialist Republics, the United Kingdom of Great Britain and Northern Ireland, and the United States of America).[12] The United States was instrumental in helping broker the final treaty during Dwight Eisenhower's tenure. Still, it was primarily shaped by the dozen or so countries that had been active near the Antarctic territory.

Eventually, they reached central agreements, including freedom of scientific research, a peaceful approach, etc. With the Cold War Era as the setting, the mandate that the continent not be used for making war—its demilitarization—was critical. It also specified regulations of Antarctic mineral resource activities, outlined in 1988, in addition to specific guidelines on marine pollution, environmental impact, and waste management. This was to ensure no dumping at the shoreline, which was common practice worldwide well into the 1990s.

In the late 1940s, multiple nations claimed territory in the Antarctic Peninsula. The debates were getting contentious, with movement toward open conflict. At the same time, some countries were fearful that the region would become a pawn in the Cold War developing between the U.S. and the Soviet Union, both of which had significant Antarctic interests. Diplomats began to talk seriously about a treaty governing Antarctic affairs. We are at a similar point today with outer space. That treaty also had been long overdue. The accord marked a significant turning point because it was concrete about the area it governs, without ambiguity. The treaty system defines Antarctica as all the land and ice shelves south of 60°S latitude.

And of course, protecting the Antarctic environment and ecosystems is in the greater interest. This element was declared outright by the treaty nations in the preamble to the historic Protocol on Environmental Protection to the Antarctic Treaty, signed in Madrid in 1991 and taking effect in 1998. The "Madrid Protocol," as it became known, came to fruition due to the agreement on the global importance of the Antarctic environment and the need to protect it. Even after the treaty was signed, the Antarctic nations recognized the importance of conservation measures in the Antarctic by initiating the Agreed Measures for the Conservation of Antarctic Fauna and Flora in 1964, to protect native animals and plants and set up protected areas.

This was a broad agreement in principle and philosophy and was made more specific in later years. In 1978, the Convention for the Conservation of Antarctic Seals provided a means for regulating commercial sealing. Should it ever resume, it fully protected three Antarctic seal species and established catch limit mechanisms for another three. In 1989, Australia decided that, instead of ratifying a planned regime to control mineral resource activity in Antarctica, it would work with the Antarctic Treaty countries for complete environmental protection for the continent and its surrounding seas. This is important because this illustrates how the example and choice of one nation helped lay the foundation for an exceptional journey of international cooperation and led to the development of the Madrid Protocol.

The Madrid Protocol was a landmark agreement because it placed an indefinite ban on mining or mineral resource activity. It designates the Antarctic as a natural reserve devoted to peace and science, with multinational, codified environmental standards. It also created a new system of protected areas, guidelines for the conservation of flora and fauna, as well as management and disposal of waste. The Antarctic Treaty and its related agreements, which have since expanded to include 50 Antarctic Treaty nations —about two-thirds of the world's population—have become representative of the interests and views of a significant number of people, and it has worked well for decades.

This level of detail may not be feasible for the Outer Space Treaty. Still, it illustrates the need for an active and progressive movement to clarify how we operate and engage in this new world of possibility.[13]

In 2017, Luxembourg established a legal framework securing private rights over resources mined in space.[14] Japan and UAE have also done the same.[15]

There need to be more explicit guidelines around how we operate in space, how different countries will collaborate, and how we tap into resources. The agreements that we would define as other countries look at what lies beyond. What if one of them wants to mine the moon or asteroids? There's much work to be done. The fundamental principle we must keep in mind is that we must continue shifting from a competitive approach—the old relics of the Space

Race between the U.S. and Russia—to a more collaborative approach. The kind of cooperation that helped us achieve the International Space Station came about through collaboration with Russia, Japan, and other nations.

Space is a Global Commons. We need to move into an era of greater symbiosis, where we realize that it's not just about collaboration but also recognizing the mutual benefits of working together. It's about working toward a mutually beneficial outcome.

As we work together, we enter into our own interstellar space. We are back to that moment between John Coltrane and Rashied Ali. We need a plan for the music we are about to make together. We're navigating space already, and we must make sense of it, define it, and tell our own story.

In our journey, to set down guidelines that will help us work together, we need agreements to help us navigate. Even if we think it's a vacuum, it's not truly one.[16] There are individual motivations and influences. Without an accord mutually agreed upon, the interstellar space economy will not be as thriving as it could be. We need to find a way to fill the space between the stars, something that gives us a plan for navigation, collaboration, and symbiosis.

SEVEN

An Interstellar Magna Carta

AS WE'VE SEEN, there are several apparent gaps in the existing treaties and agreements. Many issues still need to be solved, so they are not open to interpretation by individuals, countries, and commercial entities.

These necessary interstellar accords, what I would term our "Interstellar Magna Carta," should be formulated with care and collaboration. When I say "interstellar," I don't want to imply that we must work out these agreements with hypothetical other civilizations. We've got enough to deal with when it comes to our planet. But we need rules of engagement that will allow us to work across nation-states to navigate our journeys into outer space and beyond peacefully.

Now the original Magna Carta was drafted for self-protection by English barons in the 13th century. Two fundamental tenets have stood the test of ages and influenced other causes of democratic aspiration:

> *"No freeman shall be taken, imprisoned, disseised, outlawed, banished, or in any way destroyed, nor will We proceed against or prosecute him, except by the lawful judgment of his peers or by the law of the land."*

And

> *"To no one will We sell; to no one will We deny or delay, right or justice."*[1]

Issues of concern in our Interstellar Magna Carta must include the demilitarization of space, mining, navigation, access, and homesteading.

Space is a Global Commons where we all will participate and engage. We cannot treat it as a global arena using principles of zero-sum victory. But we must start with overarching guiding principles that provide a framework for engaging with each other.

The commons principle is not new. Humankind has been urged for many centuries to work toward a greater good and the common good. Sanatana Dharma, for instance, refers to the "eternal" or unchangeable set of obligations of all, irrespective of class, caste, or sect, and is mentioned in the Rig Veda.[2] Sanskrit uses the word "Sanatana" to denote anything that is Anadi (beginningless), Anantha (endless), and never ends, something that is eternal. Given the depth and scope of its significance, we cannot translate dharma fully into English. Its root word, "dhri," means to join, hold together, or endure. Sanatana Dharma, which means "natural law," refers to the moral precepts inherent in the universe's nature and creation.

The majority of religions teach us to prioritize the greater good. Contrary to this, modern philosophers (and economists) believed that principles of shared ownership, when applied to natural resources used collectively, would eventually cause them to be over-exploited and destroyed. Elinor Ostrom debunked this. She earned the Nobel Prize in 2009 by researching how people in tiny, local groups manage shared natural resources, like meadows, fishing seas, and woods. She demonstrated how standards are eventually developed to manage and use natural resources in economically and ecologically sustainable ways.

Even the father of modern economics, Adam Smith, did not refer to his discipline as economics. He referred to it as "moral philosophy" and saw The Theory of Moral Sentiments as being more significant than The Wealth of Nations.

He said, *"How selfish soever man may be supposed, there are evidently some principles in his nature, which interest him in the fortune of others, and render their happiness necessary to him, though he derives nothing from it except the pleasure of seeing it."*

We need to develop principles for Space as Global Commons, our Interstellar Magna Carta, with optimism. We are beyond the era of 1492, where divine jurisdiction anchors first-mover claims. We must recognize that national rights and jurisdiction are similarly insufficient and illogical to negotiate outer space. This is not about first-mover advantage. We are not sixteenth-century pioneers staking claims in the wild west, nor are we colonizers seeking zero-sum enrichment. We've evolved beyond that.

Similarly, we cannot operate as we have historically. We can use a different approach than we used for conversations on nuclear disarmament or climate change. We have learned that we cannot be successful if some nations claim superior moral ground and expect others to use a different path to progress and self-determination. If we establish joint ownership, communities and nation-states will behave responsibly.

Preliminary principles to navigate ***Space as Global Commons*** are:

1. Earth as a Unified Community - The earth is a single community. Country borders get drawn and redrawn - but for our protection and progress, our planet needs to operate as one, especially when navigating outer space.
2. Space as Global Commons - Celestial bodies and planets belong to everyone. The outer space landscape is broader and more significant than anything we have imagined or anywhere we have operated. And we have much more to learn about it. Joint ownership is the principal to anchor claims and control resources.
3. Collaborative Dialogue - All interested participants (countries) are invited and included, to collaborate and align on the principles of agreement. With inclusion, we can move beyond self-deterministic approaches that are born of necessity. Targeted dialogue to ensure multiple perspectives, even if we disagree - including those of current and potential adversaries and emerging and smaller countries.
4. Joint Governance and Oversight - International norms will oversee how we navigate discovery. We are a unified community that needs a joint course

to manage the discovery - of celestial bodies (like planets and asteroids), natural resources (like rare metals and minerals), technologies, and astrobiological markers (like nucleobases).
5. Peaceful Coexistence - The countries of the earth need to agree on space as a peaceful arena and develop norms that reinforce and allow this.
6. Navigation Safety - The safety of spacecraft, satellites, and other assets are protected by specific global agencies. We must acknowledge that conflict may arise and formulate agreements to manage these.
7. Individual Sovereignty, Agency, and Privacy - We deepen our understanding of sentience. To navigate discovery ethically, we protect individual agency, to allow physical and mental self-determination and privacy.
8. Societal Wellbeing, Diversity, and Fairness - The era of colonization allowed injustice and crime by using principles of societal and racial hierarchy, exacerbated by ignorance and insularity. Going forward, we recognize all aspects of diversity, expose biases and operate on principles of equity, so we don't make the mistakes of prior generations of explorers. As the spacefaring accelerates, certain economies, groups, and geographic regions may benefit more than others. This will broaden the gaps between urban and rural populations and between high- and lower-income countries. For common good and societal stability, we work to ensure that the benefits of discovery are shared.
9. Environmental & Planetary Protection - With the accumulation of spacecraft and residual objects,

our use of earth's orbit has exponentially increased, endangering the orbital environment. With society's increasing reliance on space-based technologies and infrastructure, we develop clear guidelines and agreements on how we will protect the earth, our natural habitat.
10. Accountability - shared accountability and integrated jurisdiction are the foundation to enable space as a global commons and allow us to operate as a unified community. We articulate clear guidelines on using international courts for space disputes and claims.

These Principles for Space as Global Commons and others will need to anchor how we resolve one of the most critical issues - the need to demilitarize space. We shouldn't use weapons in space, and space shouldn't be used as a field for military testing or dominance, but the likelihood of this happening is high and continues to rise. Space junk can create navigational hazards for astronauts and interfere with ground-based telescopes. The issue is expected to grow in the coming years as the use of internet-broadcasting mini-satellites increases. Of course, there have not been any outright space battles or military conflicts. Still, many countries have already started flexing in that direction.

In 2007, China tested an anti-satellite missile, which created space junk—debris that continues to impact other satellites. Then in 2021, Russia did the same, leaving over 1500 pieces of debris from their test. In November of that year, the International Space Station narrowly missed this debris.

On December 8th, 2022, The United Nations General Assembly passed a resolution asking countries not to conduct direct-ascent anti-satellite tests (ASATs) that create space junk. The United States spearheaded the resolution. The resolution does not ban the development or testing of ASAT systems. Still, it discourages conducting them in a way that creates space debris. Since it came from the U.N. General Assembly and not the Security Council, it is not legally binding.

Of the four countries that have conducted ASATs — the United States, China, India, and Russia — only the U.S. voted in favor of the resolution. Nine more countries abstained: India, Lao People's Democratic Republic, Madagascar, Pakistan, Serbia, Sri Lanka, Sudan, Togo, and Zimbabwe. In addition to Russia and China, other countries voting against the resolution included Belarus, Bolivia, the Central African Republic, Cuba, Iran, Nicaragua, and Syria.

As more countries see space as a potential warfighting domain, the need for multilateral discussions on space security has become more critical. These discussions aim to reduce the risk of a space confrontation that could affect all countries' ability to operate safely in this domain. Recently, there has been growing concern about the potential for conflict in space as more countries develop technologies that we could use for military purposes in this domain. This has led to increased calls for the international community to address these issues and establish rules and norms to prevent conflict in space.

As of August 2022, the U.S. Army's satellite communications mission officially transferred to the U.S. Space Force —the first time the U.S. consolidated all Department of Defense military satellite communication functions under a single military service. The move marks ongoing efforts to boost space-related efforts within the U.S. defense umbrella. It's also a move to help prevent or deter nuclear attacks through special intelligence operations providing real-time space situational awareness capabilities, a new deterrent triad. It also represents cognizance of the potential threat from space militarization, space warfare, and what it might take to defend cyberinfrastructure. It's a concern that is increasingly top of mind around the world.

Now, consider something as familiar (on earth) as mining. The resources that potentially lie on the moon or asteroids, or other planets are real, with significant value to nation-states and private corporations. That value increases for any players who participate. We also know that there have been active efforts to sample the asteroid belt. Some of this is happening under the aegis of the U.S. Commercial Space Launch Competitiveness Act, which allows the U.S. to exploit asteroids and other space resources.

Some people believe that this Act and the Artemis Accord are ways to interpret the Outer Space Treaty in ways to benefit the United States primarily. This belief displeases other global stakeholders. It's clear the issue of mining rights is quite significant and will remain so. There are implications of the potential conflict over the rights to mine, the methods allowed, etc. Already, there's a company called Asterank, which assesses asteroids and provides information about which are most viable and most profitable for mining activities. They do that based on the 6,000 asteroids that NASA tracks. Asterank found that mining just the top 10 asteroids—based on how close they are to access and how much value they might provide—could produce a profit of over $1.5 trillion. One asteroid, "Psyche," is supposed to contain over $7 quintillion of gold. That's the number seven, followed by 18 zeros. It's drawn such interest that, in 2022, if you started typing "quintillion" into a Google search, the "quintillion-dollar asteroid" would pop up immediately as a top search result.

The potential of the available material is significantly attractive if the technology develops to access it. Of course, the cost, at this point, is prohibitive. That high cost is nothing more than what we've seen overcome. Launching satellites was once cost-prohibitive to all except wealthy governments. That cost has dropped steeply. Even though the cost of space mining seems out of reach, that will change.

There's also the argument that if we can mine asteroids in outer space, we won't have to do that here on earth anymore. So it's a win for the environment that removes exploitative labor practices, economic discontent, and other woes that traditionally come with mining our planet. But we need to be wary of transporting our existing troubles elsewhere. The most interesting and vital reason for asteroid mining is that it could increase our capacity for outer space exploration. If you can find the materials and minerals you need in outer space, and if you can find a way to extract them safely and effectively, we can build outposts more easily. You don't have to ship materials from the earth so that it would expand our outer space independence and abilities. Clearly, with so much potential—good and bad—mining has to be part of an effective Interstellar Magna Carta.

Navigation and Access

Let's talk about how we get into space. So many questions....Who is allowed to launch a spacecraft, and when? Who controls access to launches? How many satellites can a company launch? Does the Federal Aviation Administration connect and coordinate with space agencies from other countries that the FAA controls now—the protocol followed with air travel now?

Within the U.S., a mishmash of regulations governed by different bodies licenses commercial space operations. Commercial remote sensing is governed by the National Oceanic and Atmospheric Administration (NOAA), a division of the Department of Commerce. The Federal Aviation Administration (FAA) oversees the safe launch and re-entry of spacecraft and rocket bodies into the atmosphere.[3] The Federal Communications Commission (FCC) controls how domestic businesses, including satellite providers, use the radio frequency spectrum. Additionally, commercial space contractors must adhere to the standards and specifications of both NASA and the DoD.

Commercial operators in the U.S. are advocating for a more unified regulatory regime, with proposals for a single system ranging from creating a sort of "clearinghouse" for prospective operators to expediting the current tri-agency regulatory procedures by providing sole control over licensing commercial activities to a "one-stop shop." Both might make the Office of Space Commerce inside the Department of Commerce the primary point of contact.

A pair of 2019 Space Policy Directives issued by the Trump administration set up Commerce as the lead agency to oversee commercial space activities.[4] It was charged with eventually establishing a new traffic management regime to ensure the safety of the ever-more crowded heavens. OSC (a unit of NOAA) was later tasked to spearhead those efforts, including taking over the job of monitoring space objects and warning commercial, civil, and foreign operators about collisions.

This shift comes with complexity. Suppose the OSC gains authority over some on-orbit activities but not others. In that case, it may create more bureaucracy and conflicting regimes, particularly if the FCC's purview expands to include debris clearance and in-orbit servicing and production.

Some stakeholders and companies are concerned that attempts to streamline the approvals could make it more onerous, impeding future innovation and investment. Some organizations, like the Space Frontier Foundation, advocate for a period of "permission-less innovation" as an alternative to a "regulatory permitting system," which would be similar to the regulatory moratorium imposed by the Clinton administration during the early years of the internet.

But we must remember that these developments are unfolding at a very rapid pace. The DoD and the Department of Commerce are discussing how they will implement space traffic management. A civilian agency will likely manage routine traffic, and the DoD will handle more high-profile issues. Regarding communication, the Department of Commerce will work with satellite operators to prevent collisions.

Crucial questions regarding data sharing concerns, accurate data measurement, data maintenance, services provided, and a more elaborate communication system are to be announced later.

Elon Musk plans many satellite launches in the next couple of years. Coordinating space objects is critical, as are clear guidelines on launching and navigating without running into one another.

Following the example of the Antarctic Treaty, a revised Outer Space Treaty also needs to consider the designation of particular management areas or protected zones to mitigate the impact in advance. For example, what do we do with the footprint of Neil Armstrong on the moon in the moon dust? Should we designate particular scientific areas? What do we do about the United States flag on the moon's surface? Next, we should consider developing a comprehensive environmental protection protocol to outline our procedural approaches, which might have an impact due to direct or indirect contamination. We also need to establish codes of conduct - each appropriate for different celestial bodies and environments, including subsurface and orbital environments. These will need elaboration on how these may apply to various categories of activities and other sectors, from scientific and public activities to commercial and industrial.

Finally, there will always be the question of ownership. The issue of claims and property boundaries extends from the world we know to the world beyond. Right now, the Outer Space Treaty (OTS) says nation-states can't lay claim to the moon or other cosmic bodies, but is this enforceable? The treaty is also silent on the rights of other entities - individuals or corporations, probably because the

OTS was drafted in 1967 when private space exploration was not on our radar. This is particularly critical to address soon, as private spacefaring increases, and clear ownership rights can enable enhanced economic activity.

So, when someone does land on the moon and builds a lunar base, what body challenges their rights? Who is the authority of those kinds of activity? We must begin laying that groundwork and do so with some urgency. Clear rules of engagement will be paramount. An Interstellar Magna Carta will provide the philosophical framework for cooperation and, hopefully, set the optimal tone of stewardship.

EIGHT

Working and Leading in the Spacefaring Economy

YOU MIGHT THINK that the absorption of innovation into our societies or cultures depends on a central agency—some guiding authority—to propel innovation. In the context of space exploration, people might argue that, in the absence of an active space agency that has built out the messaging around the need for such exploration and the vision surrounding it, it would be difficult to understand why organizations have developed an interest in space.

In the absence of a central team to demonstrate the applicability of space technologies and space exploration readiness, you could posit that the innovations may not happen. But we all know that's not true.

The reality is that, in the absence of a central agency, we still see the creation of a culture of innovation. An ecosystem that allows for ongoing innovation and fuels it (and in a more organic fashion than might otherwise have taken place). This doesn't necessarily mean that organizations should not invest and focus efforts on building an organization that develops space exploration innovations for the benefit of a particular nation-state. I'm saying that everything doesn't need to be perfect. It doesn't have to happen sequentially. It won't. Organizations have to be cognizant and build receptivity toward opportunities. What we need are more voices in the room. We need more participation in the conversation.

In the U.S. and the European Union, national space agencies have played a pivotal role in developing space technology that has parlayed into goods and services.[1] That has generated demand for conversation and investment in exploring space. But every country has a limited amount of luxury. Maybe they don't need a lot to play. Think about New Zealand, a small island country with outsized activity in the space sector through the New Zealand Space Agency. There, the leadership rose from the commercial sector—entrepreneurs who paved the way. India is another country with a very active space agency, which has developed some pivotal work. The Indian Space and Rocket Organisation (ISRO) has orchestrated missions to the moon and Mars with incredible pace and cost efficiencies.

Interestingly, more organic development is almost happening parallel to efforts to provide a centralizing force.

The complete absorption of technology mostly depends on a skilled labor pool. It's more than technology or space experts who will be required to bring us to the next level of space exploration. Organizations and nations need broad intellectual bandwidth and horsepower to harness the level of innovation that space exploration provides. The pipeline of talent is pivotal. I believe that's part of the reason why India has seen so much success in terms of accelerating the pace of commercial space activity. The same happened for New Zealand, where Rocket Lab, a public aerospace manufacturer and launch service provider, actually invested in the country's academic programs, specifically around space and STEM fields. NASA does the same in the U.S.

What we've been observing over the past few decades is the emergence of transnational collaboration beyond competition. Hopefully, a new world order will take shape as colonization's effects fade away.

If you reflect on this for a second, this has accelerated after World War II. Many more countries sought and gained sovereignty.

Private enterprise is a different story. In this context, businesses and corporations expand to straddle multiple countries, ruled by various governments and regulations as they grow and become, in some instances, more powerful than the nation-states themselves. Think of Walmart, Amazon, and ExxonMobil; some of these corporations have more significant revenue and market capitalization than the GDP of many countries.

That rise of transnational corporations set the stage for space exploration efforts that began a few decades later.[2] As we entered a new century, we saw greater stability of technology and its ability to unify  corporations, track regulatory requirements, and allow corporations to operate more consistently across the world.

We've seen the ability to balance standardization with local culture and customization. This can be seen even in how Coca-Cola or Cadbury tailor their formulae to the tastes of different countries or how Mcdonalds' adapts menus.

Those capacities become more evident and developed as more companies balance the homogeneity of goods and services with the heterogeneity of cultures. The same thing happens within organizations. More and more companies are adopting single, unifying principles. Think of the Nike model. They use the same branding, company name, and logo but execute manufacturing and design differently worldwide. Similarly, we see knowledge- and information-centric companies and consulting firms discover new ways to be active locally but operate globally.

Some of the world's most successful accounting firms operate through some variation of this model. They have a country-specific organization, but each acts under one centralized global umbrella. All align with overall goals, vision, and guidelines. So that model of increasing decentralization requires us to think about units of operation, the most autonomous single unit of operation, and how it functions to serve and operate within a specific country.

These organizations are a conglomeration of individual partners at the most basic level. Each partner is a holder of profit and loss and is tracked in terms of how they contribute. You can visualize this as a web of entities forming a global, transnational company. The unit of operation is that country entity; however, in each of these organizations, the cost center, and profit center lie with the partners. The fundamental unit of operations is those individual partners who have equity and act as profit and loss centers.

Now, these organizations form and operate based on skills and how they access different skills. In consulting firms, people come together based on skills they may have, and they are selected for projects. They work together on specific outcomes related to a client or a particular problem. When that's done, they disband and work on another project. So the fluidity of how the essential elements of an organization come together and disperse plays a role too.

The web of individual country entities has a stack of individual partners. These partners mobilize a pyramid of individuals who come together, coalesce, and move apart on needed projects, almost like an amoeba coming together and then shifting its shape again. In this model, skills are the force of attraction to bring people together to help an organization achieve its goals.

This approach has existed and even flourished historically. Think of offshore drilling. Those organizations bring engineers, managers, and skilled workers together for a particular project. They work together, they disband, and the process starts over again. This amoeba approach has simply expanded to other industries. It's part of the changing nature of organizational elements, fueled by changes in skills. One way to think about it is how different atoms come together to form a particular compound. Different product objectives can be achieved by blending different skill sets and changing the formula. The fluidity of organizations will continue to increase because that fluidity allows them to become more resilient, more flexible, and more responsive to what customers need and what the organization itself needs.

In the skills-based talent ecosystem I discussed in my previous book, we have an ecosystem of gig, contingent, temporary, and seasonal workers, who all form part of that extended ecosystem of an organization. The infrastructure becomes more complex but also much more disaggregated and spread out. The core of permanent employees becomes smaller, and concerns over proper vetting and handling of documents back and forth present essential questions about our technologies and processes. But it's all in motion. When we look forward, we'll see the permanence of the trend toward remote work and this in space exploration as well—plus, we will all be moving toward what I call the octopus model.

Yes, I'm taking you under the sea once more.

NINE

Transnational Corporations and Interstellar Ones

AN OCTOPUS, as we touched on before, has nine brains, one in the middle and one in each of its arms. What's interesting about that concept, especially with octopi being sentient beings, in my opinion, is that each of those arms can regenerate. Each arm, by itself, is entirely autonomous. It has a brain. It can create itself anew.

We're heading to this model of organizational development. Organizations still need to have a central vision and mission. Still, they will incorporate a satellite ecosystem, each with many components that engage 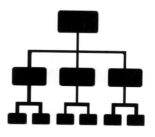 with a specific environment and are fluid enough to adapt, regenerate, and act autonomously.

Another vital piece of the puzzle here, when it comes to new organizational models, is the DAO: distributed autonomous organization. These organizations, which use smart contracts and focus on decentralization and democratization, are an outcome of the blockchain environment.

Let's shift our focus to TransNational Corporations or TNCs. A term sometimes used interchangeably with multinational corporations. There are some slight differences, but we'll talk about transnational corporations for our discussion because the United Nations Conference on Trade and Development has developed a body of thought and policy around TNCs.[1] They define TNCs as incorporated enterprises that could be parent enterprises with foreign affiliates.

This means a TNC, as defined by Trade and Development, has assets outside its home country and controls them by owning a specific equity or capital stake. TNCs typically operate in many countries, including production and service facilities. The reason I bring any of this up is that they operate in a global context. As with space exploration, the guidelines fall under cross-border jurisdictions. Of course, those corporations will have many layers of rules to follow: regional, home country, and any other country where they own or operate property.

TNCs date back to the East India Company, founded on the last day of 1600. It was one of the earliest transnational corporations that grew from trading in the Indian Ocean region to controlling the economies where it traded. Their unchecked exploitative growth is a study of unregulated commerce (and the story for another book). Jump ahead a few hundred years, and this drive to operate across national lines is a reality of our corporate world, albeit with regulation.

For the past few decades, it's been par for the course to have companies operating outside national borders as a response to several different needs. For starters, they want to grow their revenues, so they must expand into emerging markets to sell goods and services.

They also want to find avenues to trim expenses, especially for talent. Over the years, we've seen plenty of outsourcing and offshoring operations to various countries to reduce costs. How many organizations have moved transactional operations offshore? Think of the sheer number of call centers outside a company's home country. Low-cost locations can allow corporations to manufacture products at a better price point—for more significant profit margins when those products sell.

Outside of both of those catalysts for global expansion, there are several reasons a corporation might choose to stretch its footprint across international lines. Another location may be closer to the source of raw materials a company needs. Or they may decide to manufacture source goods in a location that is more conducive to acquiring new technology or near research facilities. Tax breaks or Export Processing Zones might play a role, too. Those advantages often become a sufficient reason to operate transnationally, especially when the move—or growth—allows them to be more competitive in their industries.

For some TNCs, it's also a way of hedging bets. This hedging strategy can help them balance sources of revenue and cost management. It can reduce geopolitical risks by putting its "eggs" in more than one basket. Over the years, there's been a lot of debate around the nature of TNCs and their growing economic power. So much so that The United Nations has an organization - the United Nations Centre on Transnational Corporations (UNCTC) which acts as a focal point for these organizations, which have come to comprise over 75 percent of all global trade. Transnational companies like Apple, Microsoft, Nestlé, Shell Global, and Walmart are now richer and more powerful than many nation-states. One hundred fifty-seven of the world's 200 wealthiest entities are corporations, not governments.

Similarly, out of the wealthiest 100 entities in the world, 69 are transnational corporations, not countries. And the top 10 corporate revenues exceed $3 trillion. It's no wonder that companies, instead of countries, are becoming leaders in our spacefaring ventures.

This raises another concern as we look ahead and consider agreements to pave our way—the balance of power is not just maintained among nations alone any longer. The power equation is nothing like it was in centuries past. Now we have transnational corporations (TNCs) that negotiate for lower wages, better tax rates for themselves, and greater influence in global affairs than sovereign nations. Apple's valuation at over $2 trillion in 2021 makes it larger than 96 percent of all the world's economies measured by GDP. Only seven countries on our planet have a larger economy than Apple!

The dominance that TNCs wield, even just considering sheer size, is not about power alone but resources and responsibility. We must be conscious of them as part of our geopolitical landscape and as we consider charting a new direction in this field of space exploration. We've seen debates about how these corporations should "show up" worldwide, from sustainability efforts to corporate social responsibilities. These are similar questions to whether space exploration should ultimately lead to the greater good for all humankind. It's newer territory, and the ethical questions about moral and social obligations are fair game.

That's not to say we haven't seen progress due to the spread of TNCs. In many instances, they have helped developing countries adopt technologies, industrialize, and engage in new ways in the knowledge economy of the third industrial revolution. But on the other hand, there are many criticisms about how they operate and what roles they should play.

Some TNCs have been negligent in their labor practices, putting profits over people. Some are compliant with environmental laws, and others take advantage of opportunities to use other countries as dumping grounds. Some of these are concerns we see playing out in the Congo, where, as of 2020, Chinese-backed companies owned or had a financial stake in 15 of Congo's 19 cobalt-producing mines. How much of that mining of the nation's metals and minerals—increasingly in demand as the world weans itself off fossil fuels—actually benefits the Congolese?

We have to explore the expansion of TNCs simply because they are important stakeholders in this equation for a spacefaring economy. The formula we need to organize to ensure safe, equitable, and productive space exploration will, inevitably, include TNCs. It's just the world we live in today and part of the universe we will continue to explore tomorrow.

TEN

Building the Spacefaring Company - DAOs & Blockchain

IF YOU'RE NOT a fan of the Wu-Tang Clan—or even if you are—you might scratch your head about the decision to pay $4 million for the sole existing copy of the group's album "Once Upon a Time in Shaolin." In the summer of 2021, some fans came together to buy this album and pooled their money. And the way they did it was by using a DAO: a decentralized autonomous organization.

PleaserDAO stepped up to buy this album, one of several collectibles the group of decentralized finance leaders and early NFT adopters have purchased. They claim to be pioneers, setting a precedent for bidding on unique pieces and aiming toward building an art-collecting empire. Part of their ethos is creating shared ownership of iconic pieces. Another is to provide a platform for collective experimentation that explores the

territories of community ownership, decentralized finance, and digital art (often with ownership provided through NFTs or non-fungible tokens). Their organization is decentralized and "trustless." The shared aspect is essential to them and core to how DAOs are composed.

Most important to understand is that these organizations are co-ops, or distributed organizations, with a shared bank account. They also play a role in how our next ventures into space will take flight and who will take the lead. DAOs come into play because of how they operate and use technology to empower decisions, manage work, and make transactions.[1] Imagine a way of organizing with others worldwide, even without knowing each other, establishing your own rules, and making decisions autonomously, encoded on a Blockchain or other emerging technology. That's what happens in DAOs. They provide a model of working, investing, creating, and playing together in an increasingly digital and global world.

This is a stark contrast to a traditional organization, where there is a central authority. Think of corporations: There's a board of directors exerting influence, and at the top is a CEO or president, followed by a management team who, together, run the organization. Decision-making is cascaded. A central governing authority manages decisions.

Any one person or entity doesn't govern DAOs. The rules for how the DAO operates and controls itself are coded in smart contracts on a blockchain and are changed only by voting or consensus. Instead of the CEO or a board of directors making big decisions, the majority of the members of a DAO vote on decisions—and everyone has an equal voice.

To provide a little clearer context for how this can happen, you need a basic understanding of the technology behind it. Smart contracts are collections of code recorded on the blockchain, a decentralized digital ledger. Commonly known versions of this are different cryptocurrencies, like Bitcoin, and other digital assets.

For DAOs, the blockchain (or the alternative technology used) acts as a backbone, keeping the structure and rules of each on the chain. A DAO's financial transactions and rules are encoded, eliminating the need to involve a third party in a financial transaction. The smart contract represents the rules of the organization and holds its storage.[2] No one can edit those without notice because DAOs are transparent and public. Currently, funding for a DAO is built mainly on crowdfunding that issues tokens. Governance is based on community, and operations are fully transparent and global. Decisions don't happen behind the closed doors of a boardroom. Despite the unknowns, many believe DAOs will disrupt traditional business structures.

In July 2021, Wyoming became the first U.S. state to recognize DAOs as legal entities. In June of 2022, the venture capital firm Andreessen Horowitz published an "Entity Selection Framework" describing organizational alternatives for DAOs, dividing its analysis into structure features and a flowchart to "guide builders and legal practitioners through the process of applying their facts and circumstances to the entity selection process".[3]

A wave of mainstream investment continues to build, meaning it may one day compete with traditional businesses and organizations.

Various organizations are working on establishing legal considerations for DAOs, and we will likely see more solidification of their utility in business.

To be fully functional, DAOs need a set of rules everyone agrees to follow, a funding source like tokens that the organization can spend to reward certain activities, and voting rights for establishing operating rules. Also, and most importantly, they need a secure structure that allows investors to configure the organization. Many use cases have already been identified organically. Some of today's Internet users and up-and-coming generations are seeking ways to start social organizations, looking for answers to questions like this: *How can we exchange values in a trusted environment?* Blockchain enables automated trusted transactions and value exchanges. Still, even so, people want to organize themselves in a "safe and effective way to work with like-minded folks around the globe," according to Ethereum, an open-source blockchain with smart contract functionality.

In its most basic form, a DAO is a new way for a group of people to aim for a large amount of money and organize energy at any project, whether buying a copy of the Constitution, building a new social network, or influencing an election.[4] Among the more serious-minded DAO advocates, a common theme is collective ownership. Like believers in Web3—a World Wide Web built around decentralization, blockchain, and token-based economics—DAO proponents believe we need a radically different ownership model for the next phase of the Internet. DAOs, proponents say, could allow us to build a new set of organizations and platforms owned by users, governed fairly and transparently, and native to the Internet.

You could have, for example, a DAO-governed social network where users could vote to take down certain inflammatory posts or award tokens to people who posted valuable or enlightening content. Or picture a DAO-ified version of Amazon Web Services that was run like a co-op, with members pitching in to build new features and keep the network humming.

Supporters believe DAOs have the potential to help solve some of the world's biggest problems. As technology continues to develop around smart contracts and provide the capacity to validate the credentials of people working in distant geographies across the expanses of space, DAOs will become more important to manage and understand. This swiftly emerging organizational model will shape how we head into space and how quickly.

ELEVEN

Remote Work to Interstellar Work

WHAT ARE ALL these changes in how we've done business telling us about our future, specifically the future of our pathway to space? In my book *[Block]Chain Reaction*, I covered some ways we navigate the world and how those are changing.[1] Organizations are 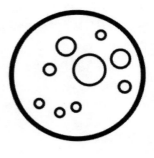 becoming increasingly decentralized, and using technologies like blockchain can help track different skill sets, verify identity, and bring together teams of people who can leverage technology to exchange credentials, work together, and get paid. Technology enables these distributed or disaggregated organizations. Co-ops are versions of DAOs. Members come together, vote, and make decisions collectively. That's what DAOs do, except they also use blockchain technology and may use a different kind of technology in the future.

And talent, as we know, is critical to any economy. We will continue to see a shift toward more employees demanding control over their work. We will see more employees wanting to own a piece of the organization where they work. Of course, this is not a new concept—it happens in co-ops and corporations with equity sharing and stock plans for employees.

The principle is that employees with a stake in an organization are more likely to be vested in its success. They are also more interested in business outcomes due to their work. That's also why CEOs and other company leaders often hold a significant portion of stock equity. This same principle applies to how we think about the next evolution of organizations, especially in a spacefaring economy.

Employees and gig workers will increasingly want to participate in an organization's decision-making. DAOs allow that. Unions have done that. Now unions present the face of disaggregated employees coming together to put their vote forward and have a group say in decisions that affect their work and working environment. We see this a lot in Europe, especially with active work councils. This was the case in the U.S. as well. The scope of that influence has changed, but unions still represent a way by which the voice of the employee affects business environments, decisions, and outcomes.

When you pull these threads together—the historical concept of co-ops and unions, which have existed or developed into employee equity in organizations and other benefits—you can see where we are headed. When employee decision-making and gig work become more interested in the distribution of governance through networks like the octopus model we discussed, we'll see what I call **"Diffused Organizations."**

These present a variety of physical locations, stakeholder groups, and distributed decision-making, all of which are fueled by technology. In a diffused organization, it will be possible to conduct remote work because a blockchain validates skills and identity. Other technologies will validate sentience, so you know you're working with real people. Decisions also can be shared through a blockchain. This is the organization of the future: more physically dispersed across geographies, more porous and permeable, and more complex and multi-disciplinary. However, there are several linchpins to making this type of organization viable.

The first is the practical speed of communications for interplanetary work. Right now, radio and light signals from earth take eight to fourteen minutes to travel to Mars (and four years to reach Proxima Centauri, the nearest star). Given the distances we are considering and the distance between Earth and Mars, we will need new, undefined ways to communicate.

Future applications and developments of quantum computing might resolve these issues.[2] NASA scientists are building a quantum laser system to relay information to

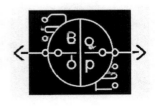

the International Space Station, using entangled photons of light to transfer data over large distances without loss. Scientists at Fermi and Argonne labs are also working on "entanglement swapping" on their journey to build what they hope will be the quantum internet.[3]

To make work *work* across physical locations, success will depend on the ability to reliably connect and work together without being together, with a shared purpose. To make the concept of a porous and permeable organization stick, you will need robust data, analytics, and precision to align external technology, gig work, subcontracting, and the aggregation of services. Success will depend significantly on the rapid absorption and diffusion of innovation and the codification of ideas. Otherwise, you'll lose innovations when a temporary team or a gig worker moves on. Finally, to navigate a more complex landscape across disciplines, we must embrace the disaggregation of skills and become skilled at assembling powerful multi-disciplinary teams to break through challenging problems.

On that note, work is also becoming more project- and outcome-oriented. Teams work together to deliver desired outcomes, and success will come only when there are shared rules of engagement and clarity of responsibilities.

Think of consulting organizations that help organizations hammer out deals or scrum teams working furiously to solve problems.

Rapidly changing technologies and markets will eventually force our hand in most cases. Organizations will need to form and reform solutions in an increasingly agile and nimble way. That means we must be willing to relearn and adopt each innovation as it emerges and balance what stays and goes each time. As I said before, we have a lot ahead of us. New models of organizing our approach to developing technologies, like space tech, are taking shape around us.

Let's think about how the Internet started. Small teams of people that trusted each other were exchanging information in closed networks—intranets. These internets were connected, became more open, and eventually formed what we now know as the World Wide Web. Over time, there was more acceptance of common standards and protocols and more trust surrounding how information would be exchanged and used.

What else happened as the Internet evolved? More and more companies started harnessing the Internet to provide different utilities. More importantly, they were using the convergence of physical access to technology with new tools to make it easier to access the technology, the World Wide Web.

What do I mean by this? When we first started working on the Internet, most of us went to a computer center or even an Internet cafe to access dial-up Internet using a modem. As technology became more ubiquitous, there was a shift: you could buy a computer. This progression, of course, is linked to your own economic or developmental situation. It also has to do with the diffusion or adoption of technology.

In one of my previous jobs, back in the '90s, we helped companies help their employees buy PCs. By that point, people were accessing the Internet in different ways. With the advent of smartphones, that access increased dramatically. Prices dropped, and people started having personal computers at home. We held the World Wide Web in our hands through our phones. This led to the rise and creation of apps and ways to provide and package services to cater to almost every need.

This platform was a decentralized web, but over time, these individual companies coalesced or grew to become mammoth organizations like Apple and Facebook that offer a centralized service made functional through decentralized work.

Now, what happens in this situation is that, as we see more scaling up of certain channels, it can begin to eliminate the ability of others to put their services or innovations in front of customers or society at large. So, if you're creating an app, you need to have an app that works on the Apple operating system. There is a standard and barrier of entry, which was more of a function of how Apple's smartphones and platforms have scaled and succeeded. The same thing is true for Android phones. If you have an app that you want people to purchase or use, you have to make sure they work on those two platforms. Those kinds of barriers can create challenges with how we diffuse innovation.

The emergence of blockchain technology, quantum computing, and post-quantum data securitization will allow us to be more dispersed in how we communicate and with whom we exchange information.

In some ways, it will take us back to the old days of the World Wide Web, where there was more of a distributed approach to how people engaged with one another. What we see in distributed autonomous organizations is that you can have groups of people who can make collective decisions by using blockchain technology. They can exchange information and vote in a democratic fashion, transforming the nature of that organization's governance. More and more, its structure is becoming flatter. It changes the dynamic. It doesn't necessarily require a central directive in the sense that we're used to seeing in corporate networks.

To that extent, we must think about mechanisms to create meaningful integration and tools that help bring people together around a common goal—especially when they can't be together in physical space.

What we see in the finance industry is something similar to a co-op version of banks, or we see it in trading with organizations like Robin Hood. It's the same for hotels, where franchisees and others have a voice in how businesses are conducted and in power grids, as more individuals will be able to have their solar power grids or even their own mini nuclear reactors. Even power grids can be decentralized.

Of course, we already see the content on Linkedin, TikTok, and Instagram stemming from a community of users. YouTube follows a similar decentralized model, where viewership correlates to revenue or endorsements.

We are going to see that kind of distribution continue across a multitude of industries. With the rise of 3-D printing, the decentralization of manufacturing will follow. NASA has granted a 3D printing and construction company, ICON, to develop Project Olympus. The project aims to use regolith from the moon or other planets to build roads, landing pads, and even homes for astronauts. 3D printing technology for construction on other planets could significantly reduce the amount of materials that need to be transported from the earth, making space exploration more feasible and sustainable. Companies that propose to do construction in space will:

1. Use hardware and software in a lunar gravity simulation flight to test their capabilities for use in space
2. Utilize samples of lunar regolith brought back from Apollo missions and various regolith

simulants to study their behavior in simulated lunar gravity.

These studies will help inform future approaches to lunar construction, including developing critical infrastructure like landing pads, blast shields, and roads. This technology could be crucial in establishing the infrastructure necessary for a sustainable lunar economy and longer-term habitation on the moon.

We will see widespread manufacturing disruption in addition to changes and decentralization in almost every arena of business and industry. Anyone might be able to print out paper clips—or a bicycle tire tube if needed. This new structure, which feels on the surface to be so unpredictable and challenging to manage, is the environment that will lead us to our new future.

TWELVE

Leaders in the Spacefaring Age

WHEN I TALK ABOUT DEMOCRATIZATION, I don't just mean in terms of countries. It's also not exclusive to corporations. Think of who our arbitrators of style are today. It's not necessarily *Vogue* alone anymore. It's  influencers on TikTok and Instagram. Who sets the tone? Regarding fashion, social media (and the masses, not Wintour) determines many trends. Social media is also a connector for political movements and activism, a new avenue for grassroots movements to congeal. People with no traditional social standing or formal leadership training can now become a voice for a movement.

We're also seeing knowledge democratization. With the rise of the Internet and remote work, especially during the pandemic, we've seen more significant opportunities and freedom regarding where people work.

People can be anywhere and still contribute to the global economy more than ever. And they don't necessarily have to have a college degree from Harvard to be successful or well-reputed. To that extent, we're seeing the democratization of knowledge sources. So much more information and expertise are coming to the forefront.

Then there's the world of ideas. We're long past the days when all innovations came exclusively from a handful of companies or countries. Innovation emerges everywhere. We've seen this with Big Pharma. They observed, years ago, that innovation in drug discovery was coming from smaller companies. Other industries have followed suit. Breakthrough innovations can come from the smallest startups, and startup culture is especially fruitful.

These changes will only increase as technology, and social media become more ubiquitous across the global economy. The traditional model is being deconstructed. Identifying top leadership requires an acknowledgment of bias—whether that's against women, minorities, transgender people, members of the LGBTQ+ community, or others who may not look or dress or sound like the cultural context we were raised to view as mainstream. Efforts to minimize bias will help us as a global community identify more leaders. We will see more women, transgender individuals, and many others who once lacked opportunities taking the helm in leadership positions. Those people formerly limited by social constructs or biases now have a platform. We will see continued shifts and breakdowns in social hierarchies and structures.

In countries like India, the caste hierarchy is being slowly eroded, and there's far more social mobility. The need to dismantle all of that—from the ostracization of specific populations to cultural structures that held many individuals in place—is a conversation taking place from the ground up in communities and countries worldwide.

With that backdrop, though, and the increasing access to power, we've also seen global autocrats' rise, whether in the Philippines, India, or the United States. We've seen more and more leaders come to power and be elected democratically who, ironically, are not pro-democracy or are more interested in building a tenure of power. That's also something we need to be mindful of, especially in the globalization of space exploration, because those power struggles are forcing change that will affect all of us.

The values of leadership are shifting too. We need people who can help us move away from ownership as a motivator, focus more on how we collaborate, and ensure we're creating a world where everyone can participate. Empathy, competency, clarity, and integrity are the characteristics that will enable our ventures into outer space. In the most basic sense, our survival depends on whether we have true servant leaders, value stewardship, and seek leaders who build progress for the next generation.

Let's circle back to the concerns over the militarization of space-an arena with potentially grave implications, to consider how nation-states have demonstrated leadership. Whether that leadership is good or bad is up to you to decide.

Increasingly it's becoming a key component of defense programs, as we discussed earlier. President Donald Trump set up the Space Force as a formal entity and arm of U.S. military services. But I'm not just talking about setting up bases or launching weapons in the future. The truth is that space is already used for military purposes. Satellites do play a significant role in military success. We're using Earth

Observation and data to inform intelligence and make decisions about threats and enemies. We also use those data and navigation systems to guide missiles and drones.

The role of space assets is already quite substantial. Satellites are computers, just a little higher up in the atmosphere. And, like computers, they can be hacked. Given this increasing reliance on satellite communication and data gathering, cybersecurity is becoming a greater and greater concern. On June 21, 2019, news agencies reported a significant NASA hack. Hackers were able to obtain access without authorization, steal "Mars Mission Data," and compromise "NASA's satellite dish network" using a Raspberry Pi.[1]

Cybersecurity represents significant threats of disruption to nation-states and companies, which then compels the militarization of our protection of satellite space. If bad actors hack satellites, it creates a threat that can disrupt entire economies. Things could escalate very quickly. Countermeasures to protect satellites and assets are the next step.

More nation-states are looking at militarization to protect their countries and demonstrate their leadership in space, which will continue. I mentioned the U.S. Space Force, which was started in 2019 and focuses explicitly on orbital and space defense, with a $24.5 billion budget for 2022—a $7.1 billion increase from the previous year. The U.S. Space Command will continue to closely monitor space observation, surveillance, and protection of U.S. assets. Similarly, France has a Space Command doing the same work. China has a Strategic Support Force set up in 2015 to find synergies and cohesion between everything China does in space, cyber, and electronic warfare.

Ultimately, the move is a defense strategy to test the capability of taking out other countries' satellites. So far, the U.S., Russia, China, and India have demonstrated this interest. Still, it's something we're likely to see gain traction as organizations and countries move toward trying to garner leadership in the space sector. In 2007 China conducted an anti-satellite test using a kinetic kill vehicle on one of its satellites. In 2020, India also performed a similar test targeting a low-earth orbit satellite.

Stellar Leadership—Four Critical Traits

As we look ahead to what factors will shape the future—and our movement toward **Stellar Singularity**—the one thing we can say for sure is that organizations will need to operate in higher uncertainty and risk. If we are working at interplanetary distances—operating machinery or supporting astronauts or servicing lunar bases at significant distances—there is a greater risk involved in operating across such vast distances. The risk is high. The ambiguity or uncertainty is also pretty high. The amount of control organizations can provide is going to decrease. What does this mean for the leadership roles the space age will require?

The same as we've seen emerge in digital spheres and a world increasingly reliant on technology. We need our leaders to demonstrate four critical traits:

1. Competence
2. Integrity
3. Clarity
4. Empathy

Competence

We've always admired competent, confident leaders. However, in this age of spacefaring, competence will become even more critical in two distinct ways. First, deep competence in one's chosen area of expertise will be paramount as the stakes increase and business decisions are no longer local or global but suddenly universal. And second, leaders will have to exhibit a core competence in talent management. I liken it to a Rubik's Cube of talent where leaders must fully grasp each job and project's complexity and dynamics. Being able to break down projects into skills required and matching those skill blocks to the right talent, and then bringing it all into a cohesive, functional whole is no small task. That capacity for whole-picture vision will become more critical than ever as we work in domains of expertise rather than set processes; working without an established hierarchy doesn't mean there is no room for leaders. They will have to be more agile and knowledgeable.

Integrity

Confucius said, "to put the world right…we must first cultivate our personal life; we must first set our hearts right." As a society, we've tolerated leaders who have lacked integrity, as evidenced by their retaining positions of power and influence long after others uncovered their scandals, misdeeds, and schemes. However, I sincerely hope that as we enter this new age, we raise the standard - perhaps as high as the stars - on integrity and expect more from those in leadership positions across business, government, academia, and so on. People build technology. Human intelligence, ingenuity, and integrity make all of this possible.

Clarity

Organizations and leaders need to be able to recalibrate the idea of certainty altogether. We have to hearken back to the unpredictability of singularity. Uncertainty is, most certainly, part of building our future. We will need to reassess concepts, ideas, and theories constantly as our pace of discovery continues. We need comfort with ambiguity and risk to keep an open mind and process new information. And then also we need clarity about the decisions and the paths we're choosing. That's the only way organizations and teams can move forward effectively in a very ambiguous situation. Progress will require focusing on incremental changes in the moment, providing clarity, and helping people understand the direction. We also must know that having something done completely or perfectly is only sometimes possible. As ambiguity increases in any environment, no answer is perfect or permanent. The answer keeps changing.

Empathy

At its core, empathy is about our ability to place ourselves in another's shoes and understand their experience. We can see how our actions affect others. The very best leaders of the spacefaring age will be the ones who can gently guide and inspire all of humanity to explore the wonders we have yet to discover. Think about it. From the International Space Station looking down on earth, it's easy to realize that we are all connected - living at the exact moment, breathing the same air, on the same planet orbiting around the sun! Empathetic leaders will deeply feel this and bring up questions of ethics, what it means to be human, and how our decisions in space will affect humankind for generations to come.

Covid-19 reminded us what we all know: being together physically and building connections in person is essential. However, it takes work to develop and maintain relationships and community with remote work. The latest research tells us that the best performers, in terms of teams, have used social connections during the pandemic to bolster their success. This offers essential clues on how organizations can foster deeper connectedness, despite remote or hybrid working situations. And the outcomes are favorable from a business standpoint. It's a solid investment to encourage human interaction among teammates.

THIRTEEN

The Temptations of the Cosmos

POWER AND DOMINANCE

The ground we laid and explored with Christopher Columbus is pertinent for several reasons. It lays out how humans have traditionally approached exploring. But it also sets the tone for the ways not only nations have evolved but also leadership.

Leadership has shifted in incredible ways over the past few hundred years: from gods to kings to democracies and increasingly to a much more diffused or decentralized model.

Using outer space as a source of power is not new. Elites and monarchs have used the cosmos as the source of authority. Kings have claimed that divine powers have installed them on their thrones; disrespecting that order was a direct affront to god and the heavens.

The Pharaohs claimed to be deities themselves. Claims of divine order emboldened even the Spanish exploration of the New World.

Spain secured the permission of the Catholic Church; therefore, they had the divine blessing to control or own whatever the country's explorers discovered.

Anthropomorphizing the cosmos to claim privilege is an old trick.

Religious leaders draw their power from the cosmos—from the forces that lie beyond what we can see and beyond our home planet. That power has been used to justify colonization, slavery, genocide, and much more over the past centuries and millennia. It's also been exchanged for material wealth and the power to command armies. The dilution began with the Magna Carta in Medieval times.

Then, the world moved toward broader knowledge, and a more shared power model, through democratization. Even though democracy existed in Greek city-states and the Roman Empire, it flourished most when colonization ended in the 19th and 20th centuries around the world. We heard ideas like "the people are the Republic." As democratic rights came to be respected and exercised, the diffusion of decision-making power extended from governments to individuals and their elected representatives.

Leadership is disseminated throughout, at least on some level. We see this even in the corporate world, where shareholders and decision-makers in some companies keep the balance of power in check. In the same way, the democratization of leadership in the spacefaring economy will have more and more distributed elements on this planet and, one day, on others.

This means the traditional ideal of a strong leader is losing relevance. As we become more conscious of acknowledging bias with the democratization of leadership, our leaders won't look the same or act the same as what previous generations upheld. We've seen many large technology firms headed by folks who haven't graduated college or are not certified engineers. Go back to the classic examples of the founders of Facebook or Microsoft. Our traditional definitions of leadership don't necessarily carry through. As technology becomes more and more accessible to people around the world, we will see more leadership emerge. With the use of technology to validate identity to connect people, we'll also see more grassroots leadership appear within organizations.

Using chat rooms or wikis within organizations helps build consensus, decision-making, or even new perspectives around a culture of employees who may not have formal leadership roles. The conversations that unfold in these media spaces shape an organization's culture. They become strong waves of implications or trends and, eventually - requirements.

Think of the backlash to policies that corporations might put forth—say, for working from home or the reimbursement of certain benefits in specific locations. It can and does drive change. Significant changes can stem from employees using technology and other forums to give themselves a voice and influence outcomes. Leaders—of companies, corporations, or even nations—need to understand how informal leadership shapes the dynamics of organizations. They need to realize the importance of identifying leadership at the grassroots level.

This presents a dramatic change in how we've looked at social hierarchy and structure. In many ways, it also disarms or replaces more formal hierarchies with structural systems that are more diffused and more distributed. These new systems are likely to create new social structures or ways of leading than we've ever seen.

Relatedly, what we also see with bias recognition, is that it's going to be very important for us to have an increasing cognizance of how bias might be screening out potential leaders.

What do emerging and future leaders in this environment need? As they navigate this new economy and this distributed, diffused organizational landscape, leaders must demonstrate empathy, competence, integrity, and the ability to operate in ambiguity. The central requirements for safe, coordinated efforts to explore outer space are the same.

We've seen examples of how the traditional approach can lead us astray. Consider Martin Shkreli, former hedge fund manager and now a convicted felon. Or Elizabeth Holmes, the biotech entrepreneur turned convicted criminal. Of course, brilliant and successful leaders can lack integrity, competence, and empathy. But as we assess leadership within our organizations and beyond them to our governments, leaders with true, lasting potential will need to be of a higher tier regarding ethics and empathy toward their subordinates.

US VS THEM

Stellar Singularity

Unfortunately, a demand for greater transparency and morals is not the only movement at play in our world today. As I mentioned, we're also seeing the rise of global autocrats.[1] Many countries are experiencing a shift away from democracy or have become democracies in name only. In many nations, we are seeing the rise of dictatorships or democratically elected leaders attempting to hold power illegally or by manipulating the principles of democracy. This is another force we must recognize and understand as we look toward the space economy. We will all be impacted by decisions global autocrats may make of their own accord.

According to the Freedom House Projects' 2021 "Freedom in the World" assessment, 2020 brought a sharp acceleration in the global decline of democracy.[2] By their measure, fewer than one-fifth of the world's population now live in fully free countries. This is part of a longer trend of democratic decline and rising authoritarianism underway across the globe for the past three decades.

How could we be moving so far backward in so little time? Because we are now living on a global scale with global waves of influence. Structural shifts following the Cold War have led us in this direction. Globalization brought about massive changes, as did the flourishing of capitalism worldwide and ongoing inequality, happening at the exact moment that globalization brings national or tribal identity to a top-of-mind state.

Global autocrats have latched onto concerns over unfairness—which people have cut in line in front of me—in the wake of financial crises. In many cases, people see only the rich benefiting, even while they are worse off than their parents. Globalization also brings about heightened potential for threats and attacks. We are still living in a post-9/11 world; fear is an easy avenue for power-grabbing leaders to exact control and have citizens pointing the blame at outsiders. The promise of protection, safety, and security, is compelling, even if it's built on a baseless foundation—or worse yet, on hate.

The "us versus them" dynamic is a throughline, and it's been used worldwide. We've seen leaders in Europe building a wall to keep out Muslims, leaders in Asia canceling the election of governors amidst fear-mongering and imprisoning people into concentration camps using the excuse of a "People's War on Terror."

These leaders, of course, don't live in a vacuum. They are observing strategies that work and putting them into play. We would do well to guard ourselves and future generations against this growing concern and its immediate dangers to safety, well-being, and progress worldwide.

Again, more than ever before, we need good leaders. Our adventures into space will require it if we are to do so safely.

Those four qualities—empathy, competence, clarity, and integrity—are vital, and employees or members of any group need to see them demonstrated more than ever before. People need to see how this translates to their actual working environment. Talking about those values doesn't make a difference.

Within an organization, they need to see how leaders support them. When employees perceive top managers as trustworthy, as people they can trust with their livelihoods and careers, everything aligns in a way it simply can't otherwise. Whole bodies of research demonstrate that integrity as a corporate value is associated with financial performance, the attractiveness of job offerings, retention, and much more.

Those four qualities will allow for big and small moments that bring us all together on this path toward Stellar Singularity. What we need is" Singular Leadership". This is the approach that will help us navigate into the future. The heart of this is a common understanding of the greater good and the ability to move people to see and work towards that greater good. That's going to set strong leaders apart, and ultimately, it's the characteristic that will lead us all toward success: a united vision across a broad, sweeping universe.

FOURTEEN

Conclusion

THE WORLD IS CHANGING–BUT it always has.

When Christopher Columbus set out across the ocean, the world people knew started expanding. Those times were full of excitement, full of peril, and possibility.

We now know that the universe is also expanding, as is our understanding. This new journey doesn't have to be frightening, and we don't have to be passive observers.

Spacefaring speaks to our nature as curious, sentient beings equipped with the potential to work together toward something bigger than ourselves.

Like my family's journey to Spain, opening our eyes to facts, history, and discoveries changes everything. This book will give you a glimpse of the considerations we need as we forge our way into New Space.

I hope you didn't just see the trajectory of humankind on its next great adventure. I hope that you also see yourself reflected. Even if you and I never step foot on the moon and never make the journey to Mars, these historic moments belong to all of us.

The next time you pop an address into your GPS, I hope you picture the satellites floating in orbit and the peaceful stillness of space. I hope you know that your own life is woven into this fabric of innovation and exploration.

I also hope you decide to step into the vacuums of Stellar Leadership. You can act as a steward for humankind and all sentient and intelligent beings—even if that role means being informed and engaged as we experience a new level of connectedness with the worlds beyond our own.

Our lives are better when everyone's lives are better. Let's harness space as a rising tide that lifts all boats and moves us all toward **Stellar Singularity;** guided not by fear but by the common good and optimism.

About the Author

Elizebeth Varghese is a futurist, people & technology strategist, and senior executive who leads and helps transform organizations around the world. She is also the bestselling author of *[Block]Chain Reaction: The Future of How We Live and Work.*

Elizebeth was recognized as among the top experts and inspirational leaders of 2022 by The HR Gazette. She has also been recognized as among The Outstanding Asian Americans in Business in 2021 and as a Top Global Influencer in 2022 and 2020 for HR Strategy and Analytics.

Elizebeth is a Board member and Co-Chair of the Entrepreneurship and Innovation Committee at Columbia Business School Women's Circle. She is on the Council of Advisors at the SETI Institute, which works with NASA and other space agencies to explore, understand, and explain the origin and nature of life in the universe.

Elizebeth has MBAs from both the Columbia Business School and the London Business School, as well as a master's in personnel management from the Tata Institute of Social Sciences and a bachelor's in pharmaceutical sciences from the University of Mumbai.

She lives in New York City with her husband and two daughters.

Also by Elizebeth Varghese

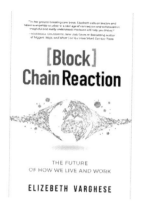

[Block]Chain Reaction: The Future of How We Live and Work, 2022

Humanity's biggest need is to connect. This will become particularly important as we use more technology and explore other planets in the future. If we move forward without keeping connection at our core, progress will be impeded and large segments of the population will be left behind. This book explains how AI in general, and blockchain in particular, can bridge our connection gap and help change the world for the better—by creating strong bonds of trust and unleashing the potential of all people, no matter where they live.

Significant Women: Leaders Reveal What Matters Most, 2021

Very few leadership books start with the simple question - What matters to you?

In *Significant Women*, you will meet corporate executives and entrepreneurs who have grappled with this question. They have led accomplished lives in the business world by revisiting their sense of significance often - sometimes by choice and oftentimes after being confronted with immense challenges on their path. Read how Elizebeth's experiences and life in India shaped her values and leadership.

Notes

Epigraph

1. Brihadaranyaka Upanishads (1.3.28). Translated from Sanskrit.

Introduction

1. 1Britannica, The Editors of Encyclopedia. "Reconquista." *Encyclopedia Britannica*, 23 Nov. 2022, https://www.britannica.com/event/Reconquista.
2. Salisbury, Joyce E. The History of Spain - Land on a Crossroad. (2017). The Great Courses.
3. Herzog, Tamar. (2015). Frontiers of Possession - Spain and Portugal in Europe and the Americas. *Harvard University Press*.
4. Pope Alexander VI (1493). Inter Caetera - Division of the undiscovered world between Spain and Portugal. *Encyclical of Pope Alexander VI*. http://www.papalencyclicals.net.
5. Britannica, T. Editors of Encyclopaedia (2022, September 21). *Treaty of Tordesillas*. Encyclopedia Britannica. https://www.britannica.com/event/Treaty-of-Tordesillas.
6. Malthus, Thomas. (1798). An Essay on the Principle of Population as it Affects the Future Improvement of Society. *J. Johnson, London*.

1. Every Company is a Space Company

1. The Future of the Space Economy. *nasdaq.com* https://www.nasdaq.com/space-economy.
2. Profile. Hermann Oberth. (2010). *NASA.gov*. https://www.nasa.gov/audience/foreducators/rocketry/home/hermann-oberth.html.
3. Williams, R. (2020). Hubble Telescope 30 Years in Orbit: Personal Reflections. *Research in Astronomy & Astrophysics*, vol. 4. Space Telescope Science Institute & University of California/Santa Cruz.
4. Mercedes López-Morales; et al (2020). Another Servicing Mission to Extend Hubble Space Telescope's Science past the Next Decade. *Astro2020 APC White Paper. arXiv:1907.04886v1* *[astro-ph.IM]* Center for Astrophysics | Harvard & Smithsonian.

5. Hubble Satellite Telescope (2022). *HubbleSite, NASA*. http://hubblesite.org/
6. Square Kilometer Array. (2022). *SKA Telescopes*. https://www.skatelescope.org/.
7. Webb Telescope. https://webbtelescope.org/.
8. Exoplanets and Water. (2022). *Hubble Site, NASA*. http://hubblesite.org/.
9. Other Global Navigation Satellite Systems (GNSS). *GPS.gov*. https://www.gps.gov/systems/gnss/.
10. Morgan Stanley. The New Space Economy. *Creating Space*. https://www.morganstanley.com/Themes/global-space-economy.
11. European Court of Auditors. Audit Preview. Maximising the benefits of the EU's space assets. January 2020. https://www.eca.europa.eu/lists/ecadocuments/ap20_01/ap_space-assets_en.pdf.
12. Editors. The Space Report 2022. Space Foundation. https://www.spacefoundation.org/2022/07/27/the-space-report-2022-q2/.
13. World Economic Forum. (2022). The space economy is booming. What benefits can it bring to Earth? *Center for New Economy and Society*. https://www.weforum.org/agenda/2022/10/space-economy-industry-benefits/#:~:text=The%20OECD%20defines%20it%20as,increase%20from%20a%20year%20earlier.
14. Joseph, Christine; Wood, Danielle. (2019). Understanding Socio-Technical Issues Affecting the Current Microgravity Research Marketplace. *2019 IEEE Aerospace Conference. Institute of Electrical and Electronics Engineers (IEEE)*.
15. Bryce Tech Reports. (2020). *2019 Global Space Economy at a Glance*. https://brycetech.com/reports.
16. Rao, Akhil and Rondina, Giacomo (2022). Open access to orbit and runaway space debris growth. *arXiv:2202.07442v1*.

2. Exploring Space Improves Life on Earth

1. Special Report No 07/2021: EU space programmes Galileo and Copernicus. https://www.eca.europa.eu/Lists/ECADocuments/SR21_07/ SR_EUs-space-assets_EN.pdf.
2. U.N. Department of Economic and Social Affairs. Sustainable Development. *United Nations*. https://sdgs.un.org/goals/goal3.
3. UNOOSA. Space Sustainability: Stakeholder Engagement Study. (2021). https://www.unoosa.org/documents/pdf/studies/Space-Sustainability-Stakeholder-Engagment-Study-Outcome-Report.pdf.
4. Woollacott, E. (2022, November 1). *How solar farms in space might beam electricity to Earth*. BBC News. Retrieved November 1, 2022,

Notes

from https://www-bbc-com.cdn.ampproject.org/c/s/www.bbc.com/news/business-62636746.amp.

5. Chen, Stephen. China's space programme will go nuclear to power future missions to the moon and Mars. (2021). https://www.scmp.com/news/china/science/article/3157213/chinas-space-programme-will-go-nuclear-power-future-missions.
6. Howell, Elizabeth. (2022). Chinese nuclear reactor proposal could power 10 International Space Stations. *Space.com*. https://www.space.com/chinese-nuclear-reactor-power-10-space-stations.
7. U.N. Principles relevant to the Use of Nuclear Power Sources in Outer Space : resolution / adopted by the General Assembly. *UN General Assembly (47th sess. : 1992-1993)*. https://digitallibrary.un.org/record/159141?ln=en.
8. 2022 COPUOS STSC – U.S. On The Use of Nuclear Power Sources In Outer Space. *U.S. MISSION UNVIE*. https://vienna.usmission.gov/2022-copuos-stsc-nuclear-power-sources-in-space/?_ga=2.64723439.1555624868.1671380811-427637587.1671380811.
9. 3-D Printed Rocket Parts and the Future of Spacecraft. *(2020). NASA*. https://www.nasa.gov/image-feature/3d-printed-rocket-parts-and-the-future-of-spacecraft.
10. Larsen, Paul B. 2021. "Is There a Legal Path to Commercial Mining on the Moon?". *University of Pittsburgh Law Review* 83 (1). https://doi.org/10.5195/lawreview.2021.821.
11. Sustainable space mining. *Nat Astron* 3, 465 (2019). https://doi.org/10.1038/s41550-019-0827-7.
12. NASA, The Artemis Accords (2020). https://www.nasa.gov/specials/artemis-accords/index.html.
13. Zell, Jeremy L., "Putting a Mine on the Moon: Creating an International Authority to Regulate Mining Rights in Outer Space" (2006). *Minnesota Journal of International Law*. 99. https://scholarship.law.umn.edu/mjil/99.
14. Jones, Andrew. (2022). China discovers new moon mineral in lunar samples. *Space.com*. https://www.space.com/china-new-lunar-mineral-chang-e.
15. Grant, John. (2021). The Moon's top layer alone has enough oxygen to sustain 8 billion people for 100,000 years. *The Conversation.com*. https://theconversation.com/the-moons-top-layer-alone-has-enough-oxygen-to-sustain-8-billion-people-for-100-000-years-170013.
16. Hoffman, A. Jeffrey; et al. (2022). Mars Oxygen ISRU Experiment (MOXIE)—Preparing for human Mars exploration. *Science Advances*. Vol 8, Issue 3. https://www.science.org/doi/abs/10.1126/sciadv.abp8636.
17. ESA. Team chosen to make first oxygen on the Moon. (2022). *European Space Agency*. https://www.esa.int/Enabling_Support/Space_Engi

neering_Technology/Team_chosen_to_make_first_oxygen_on_the_Moon.
18. Houser, Kristin. ESA plans to make the first oxygen on the moon. *Freethink.com.* https://www.freethink.com/space/lunar-regolith.

3. Private Enterprise Fueling the Next Space Age

1. Press Release. (2022). T-Mobile Takes Coverage Above and Beyond With SpaceX. *t-mobile.com.* https://www.t-mobile.com/news/un-carrier/t-mobile-takes-coverage-above-and-beyond-with-space.
2. Howell, Elizabeth. (2019). Off-Earth Manufacturing Could Help Astronauts Explore the Moon and Mars. *Space.com.* https://www.space.com/made-in-space-manufacturing-moon-mars-exploration.html.

4. Understanding Life - Here and Everywhere

1. "About the Film." The Making of My Octopus Teacher, Sea Change Project, 10 May 2021, https://seachangeproject.com/stories/the-making-of-my-octopus-teacher/.
2. Godfrey-Smith, Peter. The Mind of an Octopus. *Scientific American.* January 2017. https://www.scientificamerican.com/article/the-mind-of-an-octopus/.
3. Abbott, A. Octopus genome holds clues to uncanny intelligence. *Nature* (2015). https://doi.org/10.1038/nature.2015.18177.
4. Courage, Katherine. Harmon. Octopus Genome Reveals Secrets to Complex Intelligence. *Scientific American.* August 2015. https://www.scientificamerican.com/article/octopus-genome-reveals-secrets-to-complex-intelligence/.
5. Voinov, P.V., Call, J., Knoblich, G. *et al.* Chimpanzee Coordination and Potential Communication in a Two-touchscreen Turn-taking Game. *Sci Rep* 10, 3400 (2020). https://doi.org/10.1038/s41598-020-60307-9.
6. Kalan, A.K., Kulik, L., Arandjelovic, M. *et al.* Environmental variability supports chimpanzee behavioural diversity. *Nat Commun* 11, 4451 (2020). https://doi.org/10.1038/s41467-020-18176-3.
7. Feature. What is Kubernetes? (2017). *Redhat.* https://www.redhat.com/en/topics/containers/what-is-kubernetes.
8. Rothschild, Lynn. (2002). The Universe May Be More Habitable Than We Thought. Ad Astra Volume 14 Number 1 2002, https://space.nss.org/life-in-extreme-environments/.
9. Oba, Y., Takano, Y., Furukawa, Y. et al. Identifying the wide diversity of extraterrestrial purine and pyrimidine nucleobases in

carbonaceous meteorites. *Nat Commun* 13, 2008 (2022). doi: 10.1038/s41467-022-29612-x.
10. Y. Furukawa et al. Extraterrestrial ribose and other sugars in primitive meteorites. *Proceedings of the National Academy of Sciences*. Published online November 18, 2019. doi: 10.1073/pnas.1907169116.
11. Christopher Key Chapple, ed. Jainism Introduction. *Religions of the World and Ecology Series. Jainism and Ecology Volume*. https://fore.yale.edu/Publications/Books/Religions-World-and-Ecology-Book-Series/Jainism-Table-Contents/Jainism.
12. Shah, Umakant Premanand , Strohl, G. Ralph and Dundas, Paul. "Jainism". *Encyclopedia Britannica*, 13 Nov. 2022, https://www.britannica.
13. Jainism Simplified, Chapter 3 - Jiva (Living Beings). http://websites.umich.edu/~umjains/jainismsimplified/chapter03.html.

5. Singularity and the Spacefaring Economy

1. Article. James Clerk Maxwell - The Theory of Electromagnetic radiation. British Heritage. https://britishheritage.org/james-clerk-maxwell.
2. Niven, W.D, ed. (1890). The Scientific Papers of James Clerk Maxwell. Cambridge, The University Press.
3. Curiel, Erik, "Singularities and Black Holes", *The Stanford Encyclopedia of Philosophy* (Fall 2021 Edition), Edward N. Zalta (ed.), URL = <https://plato.stanford.edu/archives/fall2021/entries/spacetime-singularities/.
4. Kurzweil, Ray. (2005). The Singularity Is Near. Viking Penguin Books.
5. Vinge, Vernor. (1993). What is the Singularity? *Department of Mathematical Sciences, San Diego State University. Retrieved from* https://www.kurzweilai.net/the-technological-singularity.
6. REPORT / STUDY. (2019).Ethics guidelines for trustworthy AI. High-Level Expert Group on AI. European Commission. https://digital-strategy.ec.europa.eu/en/library/ethics-guidelines-trustworthy-ai.
7. Britannica, The Editors of Encyclopaedia. "interstellar medium". *Encyclopedia Britannica*, 3 Apr. 2013, https://www.britannica.com/science/interstellar-medium.
8. The Belmont Report. Ethical Principles and Guidelines for the Protection of Human Subjects of Research. The National Commission for the Protection of Human Subjects of Biomedical and Behavioral Research. April 18, 1979. https://www.hhs.gov/ohrp/sites/default/files/the-belmont-report-508c_FINAL.pdf.

Notes

9. The Belmont Report. Ethical Principles and Guidelines for the Protection of Human Subjects of Research. The National Commission for the Protection of Human Subjects of Biomedical and Behavioral Research. April 18, 1979. https://www.hhs.gov/ohrp/sites/default/files/the-belmont-report-508c_FINAL.pdf.

6. Collaboration and Competition in Space

1. Feature. Barnett, Amanda. (2022). 10 Things: Going Interstellar. *NASA, Solar System Exploration, Our Galactic Neighborhood.* https://solarsystem.nasa.gov/news/881/10-things-going-interstellar/.
2. Feature. Editor: Tony Greicius (2017). Crossing the Cosmic Void. *NASA, JPL, Exoplanets.* https://www.nasa.gov/feature/jpl/interstellar-crossing-the-cosmic-void.
3. Ibid.
4. United Nations Office for Outer Space Affairs. Committee on the Peaceful Uses of Outer Space and its Subcommittees, https://www.unoosa.org/oosa/en/ourwork/copuos/comm-subcomms.html.
5. Outer Space Treaty. (1967). Resolution Adopted by the General Assembly. 2222 (XXI). Treaty on Principles Governing the Activities of States in the Exploration and Use of Outer Space, including the Moon and Other Celestial Bodies. *United Nations Office of Outer Space Activities (UNOOSA)* https://www.unoosa.org/oosa/en/ourwork/spacelaw/treaties/outerspacetreaty.html.
6. Ibid.
7. Agreement Governing the Activities of States on the Moon and Other Celestial Bodies (The Moon Treaty, 1984), *UNOOSA.* https://www.unoosa.org/oosa/en/ourwork/spacelaw/treaties/intromoon-agreement.html.
8. Executive Order on Encouraging International Support for the Recovery and Use of Space Resources, *The White House*, April 6, 2020. https://trumpwhitehouse.archives.gov/wp-content/uploads/2020/04/Fact-Sheet-on-EO-Encouraging-International-Support-for-the-Recovery-and-Use-of-Space-Resources.pdf.
9. NASA, The Artemis Accords (2020). https://www.nasa.gov/specials/artemis-accords/index.html.
10. Feature. Shriya Yarlagadda. (2022). Economics of the Stars: The Future of Asteroid Mining and the Global Economy. *Harvard International Review.* https://hir.harvard.edu/economics-of-the-stars/.
11. History of the Antarctic Treaty. Australian Antarctic Program. https://www.antarctica.gov.au/about-antarctica/law-and-treaty/history/.

Notes

12. The Antarctic Treaty. *National Science Foundation.* https://www.nsf.gov/geo/opp/antarct/anttrty.jsp.
13. Race, Margaret S. Policies for Scientific Exploration and Environmental Protection: Comparison of the Antarctic and Outer Space Treaties Race. The SETI Institute. https://repository.si.edu/bitstream/handle/10088/16170/15.%20%20%20Race.SD.web.FINAL.pdf%20?sequence=1&isAllowed=y.
14. Boring, Nicolas. Luxembourg: Law on Use of Resources in Space Adopted. 2017. Web Page. https://www.loc.gov/item/global-legal-monitor/2017-08-22/luxembourg-law-on-use-of-resources-in-space-adopted/.
15. UAE National Space Policy-English. https://space.gov.ae/Documents/PublicationPDFFiles/UAE_National_Space_Policy_English.pdf.
16. Annual Threat Assessment of the US Intelligence Community, 2022. Office of the Director of National Intelligence, US Government. https://www.odni.gov/files/ODNI/documents/assessments/ATA-2022-Unclassified-Report.pdf.

7. An Interstellar Magna Carta

1. Magna Carta. (1297). *Edward I's Confirmation of the Charters.* https://www.archives.gov/exhibits/featured-documents/magna-carta.
2. Britannica, The Editors of Encyclopaedia. "sanatana dharma". *Encyclopedia Britannica*, 18 Jun. 2009, https://www.britannica.com/topic/sanatana-dharma. Accessed 21 December 2022.
3. Federal Aviation Administration. (2022). *United States Department of Transportation.* https://www.faa.gov/space.
4. Erwin, S. (2022, November 7). *Space force official: We need the commerce department's Space Traffic Office to be successful.* SpaceNews. Retrieved November 9, 2022, from https://spacenews.com/space-force-official-we-need-the-commerce-departments-space-traffic-office-to-be-successful/

8. Working and Leading in the Spacefaring Economy

1. Space Economy for People, Planet and Prosperity. OECD paper for the G20 Space Economy Leaders' Meeting. (2021). *Organisation for Economic Co-operation and Development's Space Forum (the OECD Space Forum) in the Directorate for Science, Technology and Innovation.* https://www.oecd.org/sti/inno/space-forum/space-economy-for-people-planet-and-prosperity.pdf.

2. Perera, A. (2022, April 20). Transnational Corporations (TNCs). *Simply Sociology.* https://simplysociology.com/transnational-corporations-tncs.html.

9. Transnational Corporations and Interstellar Ones

1. United Nations Centre on Transnational Corporations (UNCTC). (2020). *Global Civil Society Database.* https://uia.org/s/or/en/1100024712.

10. Building the Spacefaring Company - DAOs & Blockchain

1. Kaal, Wulf A., A Decentralized Autonomous Organization (DAO) of DAOs (March 6, 2021). Available at SSRN: https://ssrn.com/abstract=3799320 or http://dx.doi.org/10.2139/ssrn.3799320.
2. Szabo, Nick.(1997). The idea of smart contracts. http://www.fon.hum.uva.nl/rob/Courses/InformationInSpeech/CDROM/Literature/LOTwinterschool2006/szabo.best.vwh.net/idea.html.
3. Jenning, Miles and Kerr, David. (2022). DAO Entity Features & Entity Selection. Andreessen Horowitz. https://a16z.com/2022/05/23/dao-legal-frameworks-entity-features-selection/.
4. Single Purpose DAO. Constitution DAO. *https://www.constitutiondao.com.*

11. Remote Work to Interstellar Work

1. Varghese, Elizabeth. 2022. *Blockchain Reaction: The Future of How We Live and Work.* Soul Excellence Publishing.
2. Bub, Jeffrey. Quantum Entanglement and Information. 2020. The Stanford Encyclopedia of Philosophy, Edward N. Zalta. Metaphysics Research Lab, Stanford University. https://plato.stanford.edu/archives/sum2020/entries/qt-entangle/.
3. Awschalom, David. 2020. "From Long-distance Entanglement to Building a Nationwide Quantum Internet: Report of the DOE Quantum Internet Blueprint Workshop". United States. https://doi.org/10.2172/1638794. https://www.osti.gov/servlets/purl/1638794.

Notes

12. Leaders in the Spacefaring Age

1. NASA 2019 (Raspberry Pi) Breach OIG Report. https://oig.nasa.gov/docs/IG-19-022.pdf.

13. The Temptations of the Cosmos

1. Brownlee, J., and K. Miao. "Debate: Why Democracies Survive". *Journal of Democracy*, vol. 33, no. 4, Oct. 2022, pp. 133–49. https://www.journalofdemocracy.org/articles/debate-why-democracies-survive/.
2. Freedom House. Freedom in the World 2021: Democracy under Siege. https://freedomhouse.org/report/freedom-world/2021/democracy-under-siege.

Additional References

1. Every Company is a Space Company

1. NASA and Bryce Tech Reports. (2022). The Strategic Geography of the Solar System and Beyond. *Bryce Tech. Produced for NASA Space and Technology Mission Directorate Office of Emerging Space.* https://brycetech.com/reports/report-documents/solar_system_map.
2. Frank, Elizabeth A; et al. Why and How to Leverage the Commercial Space Sector for the Benefit of Planetary Science and Its Community. *Planetary Science and Astrobiology Decadal Survey.*
3. McCurdy, Howard and Winchell, Matthew. (2020). Cost Estimates for Sending Humans to Mars: A Continuing Conversation. *Public Policy for Innovation.* https://www.publicpolicyinnovation.com/_files/ugd/419892_28aa87ca8a4c476dbac26d128c00a458.pdf.

4. McCurdy, Howard; et al (2017). From Sailing Ships to Space Ships: An economic history of the manner in which commercial space transportation companies have utilized outside assistance historically available to emerging transportation firms. School of Public Affairs American University. https://www.publicpolicyinnovation.com/_files/ugd/419892_df8fa8c86bdf444eb6f780a4cdaadd7e.pdf.
5. Denis, Gil; et al. (2020). From new space to big space: How commercial space dream is becoming a reality. *Acta Astronautica, Volume 166, 2020, Pages 431-443, ISSN 0094-5765.* https://doi.org/10.1016/j.actaastro.2019.08.031.(https://www.sciencedirect.com/science/article/pii/S0094576519313451).

2. Exploring Space Improves Life on Earth

1. Nield, G. C. (2019). It's time for a 21st century licensing system for space. *Journal of Space Safety Engineering*, 6(4), 221-222.
2. Christensen, I., & Weeden, B. (2018). Regulatory reform is not a one-off event. *Space News*, 5.
3. EU Space Programme. *European Commission.* https://defence-industry-space.ec.europa.eu/eu-space-policy/eu-space-programme_en.
4. UNOOSA. Proposal on a Terms of Reference, Methods of Work and Workplan for the establishment of a new Working Group on the Long-term Sustainability of Outer Space Activities of the Scientific and Technical Subcommittee. *Committee on the Peaceful Uses of Outer*

Additional References

Space Scientific and Technical Subcommittee Fifty-eighth session Vienna, 19-30 April 2021. https://www.unoosa.org/documents/pdf/copuos/stsc/2021/crp/AC105_C1_2021_CRP19Rev01E.pdf.

5. UN. Report of the Committee on the Peaceful Uses of Outer Space. Sixty-fifth session (1–10 June 2022). United Nations. *General Assembly Official Records Seventy-seventh Session Supplement No. 20.*

6. Konert, Anna; et al. Legal Framework for Space Exploration. Benefits and Threats for the Earth. *Transportation Research Procedia, Volume 65, 2022, Pages 144-150, ISSN 2352-1465, https://doi.org/10.1016/j.trpro.2022.11.017.*

3. Private Enterprise Fueling the Next Space Age

1. Report. (2022). All Space Exploration Missions. *Go-Astronomy.com. https://www.go-astronomy.com/space-exploration-missions.php*

4. Understanding Life - Here and Everywhere

1. *Acaranga Sutra* 1.8.1.11–12; from *Jaina Sutras*, Part 1, The *Akaranga Sutra. The Kalpa Sutra,* trans. Hermann Jacobi (1884; New York: Dover, 1968).https://www.sacred-texts.com/jai/sbe22/index.htm.

2. Adiël A Klompmaker, Neil H Landman, Octopodoidea as predators near the end of the Mesozoic Marine Revolution, *Biological Journal of the Linnean Society*, Volume 132, Issue 4, April 2021,

Pages 894–899, https://doi.org/10.1093/biolinnean/blab001.

5. Singularity and the Spacefaring Economy

1. Lucy-Ann McFadden, Paul R. Weissman, Torrence V. Johnson. Encyclopedia of the Solar System (Second Edition), Elsevier, 2014, ISBN 9780124158450. https://doi.org/10.1016/B978-0-12-415845-0.05002-7.
2. Article. Becker, Kate. (2013). Do Blackholes Destroy Information. *PBS NOVA*. *https://www.pbs.org/wgbh/nova/article/do-black-holes-destroy-information/*.
3. Giddings, Steve. Black holes, quantum information, and the foundations of physics. *Physics Today* 66, 4, 30 (2013); https://physicstoday.scitation.org/doi/10.1063/PT.3.1946.

6. Collaboration and Competition in Space

1. Feature. Stutt, Amanda. (2020). How earth-bound mining lawyers think about space mining. *Mining.com*. https://www.mining.com/how-earth-bound-mining-lawyers-think-about-space-mining/.

9. Transnational Corporations and Interstellar Ones

1. Kordes, Marcel and Vojtovic Sergej. Transnational Corporations in the Global World

Economic Environment. *Procedia - Social and Behavioral Sciences, Volume 230, 2016,Pages 150-158,ISSN 1877-0428.* https://doi.org/10.1016/j.sbspro.2016.09.019.

12. Leaders in the Spacefaring Age

1. Hruby, Jörg & Noack, Sebastian. (2013). Global Leadership-A Narrative Synthesis. Academy of Management Proceedings. 2013. 17282-17282. 10.5465/AMBPP.2013.17282abstract.
2. Global leadership competencies: a review and discussion Jokinen, Tiina Journal of European Industrial Training; 2005.
3. Jacobs, Garry; Kiniger-Passigli, Donato and Chikvaidze, David. Global Leadership in the 21st Century. Cadmus Journal. Volume 3, Issue 6, 2019.

Made in the USA
Middletown, DE
05 November 2023